T0259827

COMPARISON OF THE LARGE-SCALE STRUCTURE OF THE GALACTIC SYSTEM WITH THAT OF OTHER STELLAR SYSTEMS

CAMBRIDGE
UNIVERSITY PRESS

University Printing House, Cambridge CB2 8BS, United Kingdom

Cambridge University Press is part of the University of Cambridge.

It furthers the University's mission by disseminating knowledge in the pursuit of education, learning and research at the highest international levels of excellence.

www.cambridge.org
Information on this title: www.cambridge.org/9781107511613

© Cambridge University Press 1958

First published 1958
First paperback edition 2015

A catalogue record for this publication is available from the British Library

ISBN 978-1-107-51161-3 Paperback

INTERNATIONAL ASTRONOMICAL UNION
SYMPOSIUM No. 5

HELD IN DUBLIN, 2 SEPTEMBER 1955

COMPARISON OF THE LARGE-SCALE STRUCTURE OF THE GALACTIC SYSTEM WITH THAT OF OTHER STELLAR SYSTEMS

EDITED BY

N. G. ROMAN

*Printed with
financial assistance from
U.N.E.S.C.O.*

CAMBRIDGE
AT THE UNIVERSITY PRESS
1958

CONTENTS

PREFACE

The field of galactic structure has developed rapidly in the past decade with important advances in several branches of optical astronomy as well as in the new discipline of radio astronomy. However, without the guidance provided by studies of other galaxies, our interpretation of the new results would have been seriously delayed. Recent progress in both of these fields was reviewed in a symposium on 'Comparison of the Large-Scale Structure of the Galactic System with that of other Stellar Systems' which was held in Dublin in September 1955 in connexion with the Ninth General Assembly of the International Astronomical Union.

Each symposium participant was requested to submit a written manuscript. Some of these were written later and contain material not presented at the symposium. In addition, the discussions and Dr Baade's introductory paper were transcribed from a tape recording of the symposium proceedings. These transcriptions were later checked and revised by the participants concerned. The unfortunate delay in publishing these collected papers in book form is partly due to the late submission of manuscripts by several of the authors.

English versions of the manuscripts were provided by the authors, but those not originally written in English were edited. The contributions by Kopylov, Vorontsov-Velyaminov and Shchegolev were not presented at the symposium but have been condensed for inclusion in this volume at the request of Dr Kukarkin.

The undersigned wishes to express her thanks to Dr A. Blaauw for his assistance in the preparation of this volume.

N. G. ROMAN, *Editor*

1. LARGE-SCALE STRUCTURE OF
SPIRAL NEBULAE

W. BAADE
Mount Wilson and Palomar Observatories, U.S.A.

For a discussion of the large-scale structure of spiral nebulae, we should like, of course, to choose a stellar system which we are certain is similar to our own. Since we know that our own Galaxy is a spiral galaxy, this narrows our choice to the spiral galaxies and, of those, we can immediately eliminate the so-called barred spirals, as we know that our galaxy is not a barred spiral. Hence, we need only consider spirals of types Sa, Sb, and Sc on Hubble's system. NGC 4594 is an Sa system with a large central spheroidal system typical of these early-type spirals. Undoubtedly, our Galaxy does not have such a large spheroidal system since it would be obvious as a large bulge which simply has not been observed. Next, M 81 is typical of the Sb spirals in which the central system has shrunk considerably. Finally, M 33 is typical of the Sc spirals in which the central system has shrunk until it actually approaches a semi-stellar point.

There are a number of reasons to believe that our own Galaxy is a spiral of type Sb. Since we are eight or nine kiloparsecs from the galactic center, the picture taken by Mr Code with the Henyey-Greenstein camera gives a first impression of how our Galaxy would look from the outside if it were seen edgewise. In this picture, the galactic center regions and the flattening in the outer parts are visible. On the basis of such a picture alone, we would classify our Galaxy as an Sb spiral. In fact, it is very similar to the spiral NGC 891, also seen edge-wise. In addition to the structural data, we could introduce other evidence; for example, the strength of the nova phenomenon. This is very strong in the Sb spirals like the Andromeda nebula and M 81 and very weak in late spirals like M 101.

For this reason, I should like to describe the Andromeda nebula as an example of the spiral structure of Sb galaxies. We have every reason to believe that something similar, both in features and arrangement, will be found in our own Galaxy. Starting with the center, we have a tiny nucleus, quite sharply defined, with the dimensions, $2''.5 \times 1''.5$. It is an elliptical structure, oriented exactly like the major axis of the whole system. A very important feature is that the spiral structure, in the form of dust clouds, reaches right into the nucleus, on both sides. The first spiral arm comes out, swirls around, and then splits into two sections. One continues around, the other shoots further out. The first part of the innermost spiral arm is very disrupted. The clouds forming it are huge flares in a very turbulent state. The next spiral arm, number two, is barely visible because it is filled in with the light of the nucleus, but it shows a remarkable feature. Along one edge, in a dust lane, the first stars appear. These stars are all blue super-giants, and as soon as we find them we also find H II regions. Farther in, neither H II regions nor stars are observed. Hence, at a distance of about 3 kpc from the nucleus, the first stars and the first H II regions are observed in an otherwise dark spiral arm. The next arm is again essentially a dust arm, but everywhere along the outer fringes of this arm are super-giants which are associated with the arm. The huge star association NGC 406 is in the next arm. In this region of the Andromeda nebula the spiral structure is very complicated. The arm branches to form a loop of cloud complexes from this arm to the next. Spiral structure is never as regular as has been thought. Hubble concentrated most of his investigations on the next arm in a region about 50′ from the center of the Andromeda nebula. In front of darker material, this arm appears bright. It is full of giant and super-giant stars, by which it is essentially marked instead of by gas and dust. H II regions are present, but only in the dark lanes or along the edge of a few large dark clouds. Although in other areas there are plenty of O- and B-stars, the H II regions are restricted to these dark areas. Hence, we must conclude that elsewhere dust has already been converted into stars, and dust and gas are essentially absent. Where the arm crosses in front of the nebula, it appears simply as a dark lane. H II regions are present throughout the interior of the lane, thus showing evidence of star formation in the center of the dust lanes. The remaining arms are essentially star arms with relatively few emission nebulae. Skipping one arm, we come to the one about $1°5$ from the center of the Andromeda nebula. It is very thin; most of the stars are blue super-giants with red ones among them. Extra-galactic nebulae shine through the arm in large numbers. There are only very few emission nebulae which are associated with O- and B-star associations. Nevertheless, the arm contains a large number of variable stars, many of them Cepheids. Going still farther out, we find the last arm at about 2° from the center of the galaxy. It is still thinner than the preceding one, but even in it an occasional emission nebula is found. Beyond this we find what are apparently remnants of former arms. For example, in a remnant 2° 14′

from the center, we find a little group of B-stars which are about 20 kpc from the nucleus of the Andromeda nebula.

The picture of the spiral structure can be summarized as follows: In the interior, we see the spiral arms essentially as dust arms, but soon, at a distance of about 3 kpc from the center, we can detect the first super-giants and the first emission regions. From there out, the spiral arms are dotted with super-giants and H II regions, but only at the outer fringes of the dust arms. The central parts of the arms still appear only as dust. By the time we reach a distance of about 7 to 8 kpc from the center, the arms appear essentially as arms formed by stars; dust and gas are still noticeable, but they become less and less conspicuous as we go to spiral arms beyond 8 kpc from the center. Beyond this, the arms are thin and primarily represented by stars. At a distance of nearly 2° from the center, far beyond what is usually considered the body of the Andromeda nebula, the outer-most arm can still be traced. Of course, this entire spiral structure is embedded in a huge disk of population II stars which extends far beyond the spiral structure. In the region between NGC 205 and the Andromeda nebula, the whole field is covered with population II, including globular clusters.

Discussion

? : Do we know the near side of the Andromeda nebula definitely?

Baade: We conclude that one side is the near side from the orientation and the structure of the absorption lanes.

Oort: How good is the evidence that there are no blue super-giants nearer than 3 kpc to the center?

Baade: The only evidence is that I have been unable to find a single H II region there, although I have made every attempt to do so. I think they would be visible even against the bright background. Also, the structure of the inner spiral arm is very disrupted and gives the impression of being very turbulent.

Lindblad: If you assume that these clouds near the center of the Andromeda galaxy lie in its plane, and try to untilt them, they become extremely elongated along the minor axis. Thus it is interesting that there is fairly good evidence that the clouds do not lie exactly in the nebular plane but are well elevated.

Baade: Yes, there is certainly evidence that the spiral structure in the central region does not lie in the plane.

2. SOME REMARKS ON MULTIPLE GALAXIES

V. A. AMBARTSUMIAN
Bjurakan Observatory, Erevan, U.S.S.R.

(Read by Mrs Masevich)

Within each cluster of galaxies, the ratio of the number of multiple galaxies to the number of single galaxies is much higher than the ratio calculated under the assumption of statistical equilibrium. This fact leads, without any supplementary assumption, to the conclusion that the components of any given multiple galaxy have a common origin. In this respect the multiple galaxies are similar to the multiple stars of our stellar system.

The study of the configurations of the multiple galaxies in Holmberg's catalogue has shown that among 132 multiple systems in this catalogue, 87 systems (65 %) form configurations of the Trapezium type, while only 27 (20 %) belong to the type usual for multiple star systems. The remaining 15 % are intermediate in type with the ratios of the largest distances between components to the smallest ones between 2·5 and 3·0.

In this respect the class of multiple galaxies differs strongly from the class of multiple stars, in which the Trapezium type configurations form only a small minority.

The Trapezium type multiple galaxies are probably unstable, at least in the cases in which the masses of the components are similar. The very existence of such configurations makes it probable that many of them have positive total energies. This means that some of the multiple galaxies represent newly formed and expanding groups, but this assumption requires some independent proof.

From the differences in the radial velocities, Δv_r, between the components of the pairs of galaxies observed by Page and others, we can find the values of $\rho(\Delta v_r)^2$, where ρ is the projected linear distance between the components. These values of $\rho(\Delta v_r)^2$ are systematically much higher for the pairs in the multiple systems with three or more components than in simple double galaxies.

However, if all the systems have negative total energies, the average value of $\rho(\Delta v_r)^2$ will be proportional to the average mass of the systems. Thus,

4

the assumption of negative total energy for all systems leads to the conclusion that the masses of the components of multiple galaxies are systematically greater (at least by multiples of two or three) than the masses of the components of double galaxies.

The only way to avoid this conclusion is to assume that some multiple galaxies have positive energies. Thus we have an indirect proof of the assumption introduced above.

The mean value of the ratio $f = M/L$, derived for multiple systems, assuming negative total energies, is much higher than the maximum value of the same ratio obtained from the rotation of individual galaxies. This may provide additional evidence in favour of positive total energies for some multiple galaxies.

If we assume the steady state of clusters of galaxies and apply the virial theorem, we obtain very high values for the ratio f. Thus we obtain $f = 2000$ for the Virgo cluster and $f = 5000$ for the Coma cluster, if we take the new value determined by Zwicky for the radius of the latter cluster (and not $f = 800$ as derived by Schwarzschild, who applied only the corrections which tend to minimize the mass and neglected the opposite corrections).

Though the evidence in favour of inter-galactic matter in clusters of galaxies is convincing, it seems improbable that the mass of this matter in a cluster can exceed many times the total mass of member galaxies. Instead, it seems that the nature of inter-galactic patches is similar to that of irregular galaxies in which the ratio f is very small. Therefore it is reasonable to assume *that some of the clusters of galaxies are systems of positive energy*. In this case the virial theorem is not applicable.

As Markarian has pointed out, the presence of a chain of bright galaxies in the dense part of the Virgo cluster is an evidence in favour of the non-steady state of this system. The chain includes the giant elliptic galaxies M 84 and M 86 and is one of the outstanding features of the cluster. In this respect there is a remarkable similarity between the irregular clusters of galaxies and the O-associations, in which chains of blue super-giants appear (for example, the Orion Belt).

If among the multiple galaxies we have systems of positive energy, it is also natural to expect the occurrence of such cases among double galaxies. A double galaxy of this type would represent a pair of mutually receding stellar systems. There is no configuration-criterion in the case of double systems. However, the question arises of the possibility of observing such pairs in the early stages when the mutual distance of the components is very small and the nucleus of one component is situated in the central part of the other.

5

Actually, we observe some such narrow pairs as radio-galaxies. It is easy to show, applying statistical considerations, that the galaxy NGC 5128 (Centaurus A) cannot be a result of the collisions of two previously independent galaxies. In the case of Cygnus A the collision hypothesis should be abandoned on the grounds of similar statistical reasoning. In the case of NGC 4486 (Virgo A) the division of the nucleus into the main mass and a jet was directly observed by Baade. Therefore, we may consider a radio galaxy as a result of the division of the nucleus of some primary system.

Discussion

Minkowski: Results of a detailed investigation of NGC 1275, reported at the Jodrell Bank Symposium, admit no other interpretation than that this system consists of two colliding galaxies.

3. A STUDY OF BRIGHT MEMBERS OF THE LARGE MAGELLANIC CLOUD

M. W. FEAST, A. D. THACKERAY AND A. J. WESSELINK

Radcliffe Observatory, Pretoria, South Africa

(This is an abstract of the paper presented by A. D. Thackeray at the symposium. The full paper is in *The Observatory*.*)

A progress report on work on (*a*) spectral types and (*b*) radial velocities of twenty-eight members of the Large Cloud (mostly $m_{pg} = 11$ or brighter) will appear shortly in *The Observatory*. The commonest spectral types found lie between B 5 and A 2. A star with $M_{pg} = -10 \cdot 1$ has slightly narrower H-lines than the galactic A 2 super-giant HD 92207. [Fe II] emission appears in four of the stars, including recent spectra of S Dor.

The radial velocities exhibit the rotation of the Cloud about an axis in position angle 75° agreeing within 5° with de Vaucouleur's minor axis derived from the outer structure.

The angular velocity appears to be constant out to a projected distance of some 3 kpc. This central body rotates with a period about 180 million years.

The angular velocities of stars and diffuse nebulae appeared to be 3·3 times that of the H I gas, as suggested by the published radio data at the time of submitting the paper. The explanation of the discrepancy was offered that the complexity of 21-cm profiles might have led to under-estimated velocities for the inner core. This explanation has been con-firmed by verbal conversations at the Dublin General Assembly of the Union.

Discussion

Morgan: Can Dr Thackeray tell us the magnitude of, say, the fifth or tenth brightest of the A super-giants which he has observed in the Magellanic Clouds in order to give an idea of the nature of the bright end of the luminosity curve as compared with individual isolated members.

Thackeray: About tenth magnitude, apparent. I should say that these magnitudes are extremely unreliable. Most of these stars are variable and, if they are faint, they tend to show emission lines of forbidden iron.

Morgan: Are these A super-giants variable as a group?

Thackeray: Many of them are.

* *Obs.* 75, 216, 1955.

7

4. ON ROTATING RING ORBITS IN GALAXIES

BERTIL LINDBLAD AND PER OLOF LINDBLAD

Stockholm Observatory, Saltsjöbaden, Sweden

If ω is the angular velocity of circular motions in a stellar system depending on the distance R from the centre, and κ is the frequency of oscillation in the radius vector for an orbit in the equatorial plane which differs slightly from a circular orbit, we may show[1] that in a coordinate system of angular speed

$$\omega - \tfrac{1}{2}\kappa$$

the orbit in question is a closed oval, in the first approximation an ellipse, with two opposite vertices and with its centre in the centre of the system. The conditions are illustrated in Fig. 1, when AA is the apsidal line of an orbit. The expression for κ is

$$\kappa = 2\sqrt{\omega(\omega - A)},$$

where A is Oort's constant of differential rotation. If the distribution of matter in the stellar system may be approximated by a sum of concentric spheroids with one and the same plane of symmetry, the density increasing with decreasing dimensions of the spheroids, we should expect $\kappa = 2\omega$ close to the centre, whereas in the outermost regions we must expect that κ approaches to ω. It is therefore likely that $\omega - \tfrac{1}{2}\kappa$ does not change appreciably within the system. Fig. 2 gives the run of κ, ω, and $\omega - \tfrac{1}{2}\kappa$ for the galactic system according to the values of ω given by Kwee, Muller, and Westerhout[2]. In this case $\omega - \tfrac{1}{2}\kappa$ changes very slowly with R. A relative orbit has a quite important property, if $\omega - \tfrac{1}{2}\kappa$ is constant in the interval of R covered by the orbit. In this case, if the centre of gravity of a cloud of free particles follows the orbit, and if the velocity dispersion is small, the particles will pursue orbits which differ only slightly from each other and from the central orbit, but on the other hand there will in general be considerable differential motions along the central orbit. This orbit is therefore the curve in space along which a cloud of free particles tends to disperse. The process is similar to the dispersion of the meteoric particles of a comet along its orbit.

If we follow a piece of matter in its motion along a slightly oblong orbit of this kind, the amount of matter per unit of the central angle θ will vary

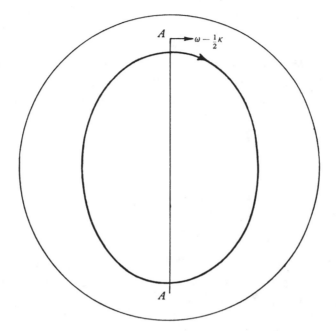

Fig. 1.

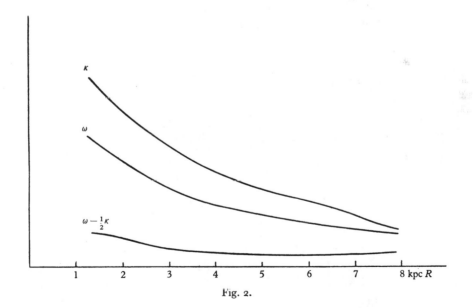

Fig. 2.

9

due to the law of areas in the motion. In the case of a perfect mixing of matter along the orbit, which is of course an ideal case, but which will be approached as time increases, there will be steady maxima of density at the vertices (major axis) and steady minima at the ends of the minor axis. Due to this circumstance there will be secular disturbances of the motions in the ring, as well as on particles outside of the ring. In order to test this effect we have made computations on the motions of ring formations in a typical stellar system.

We assume as an interpolation formula in the interval of R considered a central force of the form

$$\frac{\partial \phi}{\partial R} = -\frac{aR}{1 + bR^3}. \tag{1}$$

It is easily found that in this case $\omega - \frac{1}{2}\kappa$ is nearly constant in a region about $R = R_0$, where

$$bR_0^3 = 2 \cdot 456.$$

If we choose $R_0 = 7$ kpc, we have

$$b = 0 \cdot 00716. \tag{2}$$

The value ω_0 for $R = R_0$ may be chosen in accordance with the value derived by Kwee, Muller and Westerhout for $R = 7$ kpc in the galactic system, $\omega_0 = 0 \cdot 0323$ km/sec pc, which gives

$$a = 0 \cdot 003606. \tag{3}$$

If we take as the unit of time 10^6 years, as the unit of distance one kiloparsec, and express ω in kpc/10^6 yrs kpc, we find

$$\frac{\partial \phi}{\partial R} = -\frac{0 \cdot 003768R}{1 + 0 \cdot 00716R^3}. \tag{4}$$

The total mass of the stellar system is found to be of the order $100 \cdot 10^9$ solar masses.

At $R_0 = 7$ kpc we have

$$\omega - \frac{1}{2}\kappa = 0 \cdot 010454 \text{ kpc}/10^6 \text{ yrs kpc}. \tag{5}$$

The orbit of a particle in the central field (4) in a co-ordinate system ξ, η rotating with the angular speed (5) has been computed starting with the initial values

$$\xi = 0, \; \eta = +7 \cdot 7, \; \dot{\xi} = +0 \cdot 12285, \; \dot{\eta} = 0.$$

The resulting orbit is very nearly closed. The positions and velocities in the first quadrant (ξ and η positive) were taken to define the orbital motion at intervals of time corresponding to $1/32$ of a period. The values in the other quadrants were formed by symmetry. In this way we find the initial distribution of thirty-two points along the periphery of the ring as shown in Fig. 3.

Denoting with x_i, y_i the co-ordinates of a point i in a fixed co-ordinate system, the equations to be solved are as follows

$$
\left.
\begin{aligned}
\frac{d^2x_i}{dt^2} &= -\frac{0 \cdot 003768}{1+0 \cdot 00716R_i^3}x_i - k\sum_{j \neq i}\frac{x_i-x_j}{[(x_i-x_j)^2+(y_i-y_j)^2]^{\frac{3}{2}}}, \\
\frac{d^2y_i}{dt^2} &= -\frac{0 \cdot 003768}{1+0 \cdot 00716R_i^3}y_i - k\sum_{j \neq i}\frac{y_i-y_j}{[(x_i-x_j)^2+(y_i-y_j)^2]^{\frac{3}{2}}}, \\
R_i &= (x_i^2+y_i^2)^{\frac{1}{2}},\ i,j=1,\,2,\,\dots\,32.
\end{aligned}
\right\}
\tag{6}
$$

Choosing for each point a mass of $5 \cdot 10^7$ solar masses gives for the total mass of the ring $16 \cdot 10^8$ solar masses, i.e. between 1 and 2 % of the entire mass of the system. In this case we find

$$
k = 2 \cdot 2462 \cdot 10^{-4}.
$$

In order to compensate the increase of central force on the particles due to the attraction of the ring itself, the initial velocities of the particles have been increased by a rotational motion which compensates the increase of central force for a circular arrangement of the thirty-two points with the radius $R = 7$ kpc.

The numerical integration of equations (6) was done with the Swedish electronic computer BESK, constructed and run by the Swedish Board of Calculating Machines at Stockholm. We are very much obliged to the Board for its kindness to allow us to use the machine for the present purpose. For working purposes the equations (6) were first transformed. On account of the symmetry in distribution and velocities of the particles the equations for the first sixteen may be written:

$$
\left.
\begin{aligned}
\frac{dx_i}{dt} &= \dot{x}_i, \\[4pt]
\frac{dy_i}{dt} &= \dot{y}_i, \\[4pt]
\frac{d\dot{x}_i}{dt} &= -\frac{ax_i}{f+b(x_i^2+y_i^2)^{\frac{3}{2}}} - \frac{cx_i}{4(x_i^2+y_i^2)^{\frac{3}{2}}} - \sum_{j \neq i}\frac{c(x_i-x_j)}{[(x_i-x_j)^2+(y_i-y_j)^2]^{\frac{3}{2}}} \\
&\qquad\qquad - \sum_{j \neq i}\frac{c(x_i+x_j)}{[(x_i+x_j)^2+(y_i+y_j)^2]^{\frac{3}{2}}}. \\[4pt]
\frac{d\dot{y}_i}{dt} &= -\frac{ay_i}{f+b(x_i^2+y_i^2)^{\frac{3}{2}}} - \frac{cy_i}{4(x_i^2+y_i^2)^{\frac{3}{2}}} - \sum_{j \neq i}\frac{c(y_i-y_j)}{[(x_i-x_j)^2+(y_i-y_j)^2]^{\frac{3}{2}}} \\
&\qquad\qquad - \sum_{j \neq i}\frac{c(y_i+y_j)}{[(x_i+x_j)^2+(y_i+y_j)^2]^{\frac{3}{2}}}, \\
&\qquad\qquad\qquad\qquad i,j=1,\,2,\,\dots\,16.
\end{aligned}
\right\}
\tag{7}
$$

Keeping the numerical values of the constants in (6), the units of distance and time have been chosen so that the constants a, b, c and f satisfy the requirements of the machine. The equations (7) were then integrated by the method of Runge and Kutta with a constant full-step of 8.10^6 years.

The results obtained for the single ring of thirty-two points are shown in Figs. 3–7. The number in each figure is the time expressed in the unit

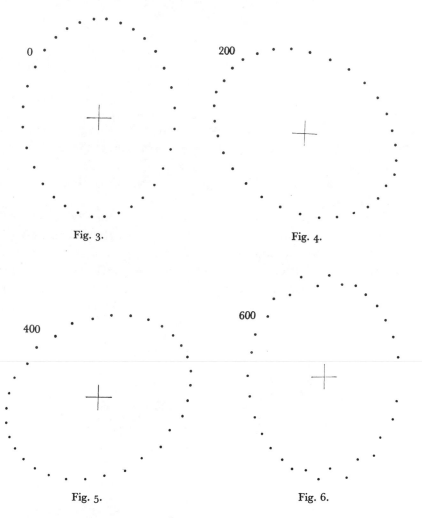

Fig. 3.   Fig. 4.

Fig. 5.   Fig. 6.

10^6 years. The disturbances are very small up to $t = 400.10^6$ years. At 600.10^6 years certain deviations from the smooth distribution occur and Fig. 7 shows that there is a tendency to break the ring into two separate ring-halves. At this time, however, certain points got so close together that the disturbing force exceeded the limits set by the machine with the present arrangements.

It may be remarked that this breaking up of the ring may be predicted analytically, as will be shown in a forthcoming paper on the present subject.

Of still greater interest is, however, the influence of the disturbing forces of a slightly oblong ring on neighbouring rings of matter, which may be assumed to have circular symmetry. In order to investigate this case we have distributed the thirty-two points on two rings, sixteen points on each. The outer ring is identical in form with the earlier one; the radius of the inner circular ring is 5·3 kpc. As before, the initial rotational velocities were adjusted to compensate the central force produced by the

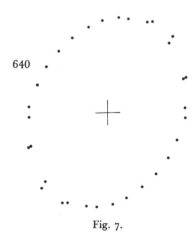

Fig. 7.

combined attraction of the rings. The results of the computations are illustrated in Figs. 8–11. In this case too the perturbations are small until a time of about 400.10⁶ years. At this time a widening and a beginning break-up of the inner ring is indicated. At 624.10⁶ years this process is well marked. The inner ring has a tendency to break up into two halves, which, moreover, tend to become mixed with the matter of the outer ring. For larger values of t, coincidences between points set a limit to the calculation with the present arrangements.

Similar computations under varied initial conditions will be continued in the future. In particular it will be tried to cover larger intervals of time in the computations.

Already the preliminary results obtained here show beyond doubt the importance of such calculations for the problem of the development of spiral structure in galaxies. The theoretical concepts underlying an attempt to explain the spiral structure along these lines may be very briefly summarized as follows:

(1) The properties of symmetry which the spiral structure displays, and the general regularity of its pattern in other respects, for instance in the barred spirals, in spite of individual features which differ from one object

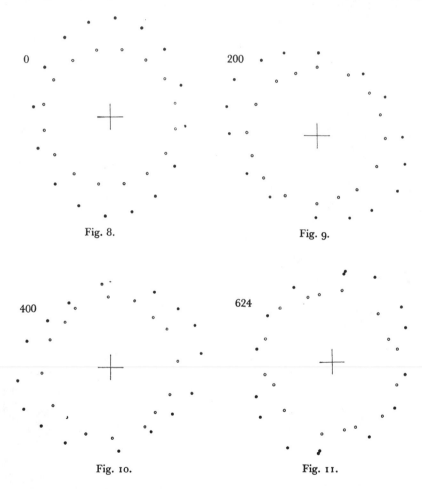

Fig. 8.   Fig. 9.

Fig. 10.   Fig. 11.

to another, appear to indicate that the main cause of the spiral phenomenon should be sought for in gravitational disturbances. Other forces may not be able to produce effects of such regularity over the entire mass of the system.

(2) The process of formation of slightly elongated rings of matter is likely to be an entirely normal process in the evolution of a system. It

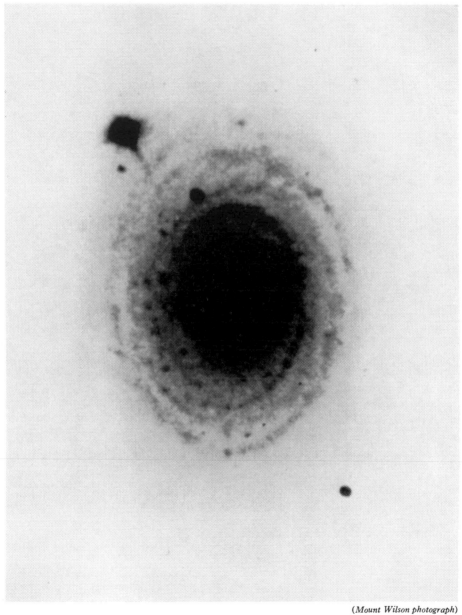

(*Mount Wilson photograph*)

Fig. 12. The spiral galaxy NGC 488.

Facing p. 15

admits of an age of the spiral structure practically equal to that of the system itself. The development of spiral structure is in principle a slow process and is due to secular disturbances of the motion, which follow immediately from the average distribution of matter in the rings. This does not exclude that certain phases in the development of the structure may proceed rather rapidly.

(3) The symmetry of the spiral pattern follows immediately from the symmetry of the rings, and needs no auxiliary hypothesis.

A type of spiral system for which a direct application seems to be possible may be represented by the nebula NGC 488 (Fig. 12). In this case we can readily imagine that the spiral pattern has originally consisted of a system of rings, which have been broken up by a symmetrical disturbing field of force. The apparent elongation of the outer rings may be at least partly real, because the inner spiral structure, which is overexposed in the picture, appears to be more nearly circular.

REFERENCES

[1] *Stockholm obs. Ann.* **18**, no. 6, 1955.
[2] Kwee, K. K., Muller, C. A. and Westerhout, G. *B.A.N.* **12**, 211, 1954.

5. THE DISTRIBUTION OF ATOMIC HYDROGEN IN THE GALAXY

H. C. VAN DE HULST

University Observatory, Leiden, Netherlands

It is my privilege to report here about the investigations of the distribution of neutral hydrogen in the Galaxy. These observations, conducted by the Netherlands Foundation for Radio Astronomy, were made at Kootwijk and reduced at the Leiden Observatory. More than a dozen persons have participated in this work during the past year. Among them are Prof. Oort and myself, but I should mention three in particular: C. A. Muller has perfected the instruments even further and has supervised the observations that were made during 24 hr a day. G. Westerhout has supervised most of the reductions at Leiden and has prepared the map and model of the outer parts of the Galaxy. M. Schmidt has unravelled the situation in the inner parts of the Galaxy, on the basis of his observations made for this purpose.

The observational material gives, after proper corrections for instrumental effects, the *intensity* as a function of galactic co-ordinates l, b, and the frequency v. We usually scan in v at fixed l and b, thus recording a line profile. As the frequency shifts must certainly be ascribed to relative velocities v of the atoms and the observer, we can convert to (l, b, v) as co-ordinates. Our final results are presented in the form of graphs giving the *density of atomic hydrogen* as a function of the three space co-ordinates (l, b, r) or (x, y, z). Obviously, certain assumptions have been made in this reduction. It seems proper to devote this talk to a summary of the main results together with a discussion of the reliability of the assumptions on which they are based.*

We may distinguish four regions of the Galaxy (R = distance to galactic centre):

(*a*) near the plane, outer parts, $R > 8$ kpc,
(*b*) near the plane, inner parts, $3 < R < 8$ kpc,
(*c*) near the plane, central parts, $R < 3$ kpc,
(*d*) up to 2 kpc from the plane, mainly local.

---

* The results are summarized in greater detail in *Radio Astronomy* (I.A.U. Symposium no. 4, 1955), papers 4, 5 and 6, by G. Westerhout and M. Schmidt, Cambridge (1957). For complete data see *B.A.N.* No. 475 (1957).

The basic assumption made in reducing observations in regions (a) and (b) is that the velocities consist of a systematic circular motion around the galactic centre combined with random, isotropic, cloud velocities. The latter are called cloud velocities as the thermal motions of the atoms are negligible. No attempt is made to see individual clouds; the instrumental dimensions are such that an average is taken over a volume of space that is several hundred parsecs in each of the three dimensions. Further, in reducing observations in (a) and (b), the assumption is made that the angular velocity $\omega(R)$ and the distribution of cloud velocities in one component $f(v)$ are known. Most of the criticisms may enter here. There may be local or systematic deviations from circular motion, or we may have the wrong $\omega(R)$. Such criticisms are good for warning us against over-interpreting the data, but at the moment the following considerations seem to indicate that no serious deviations exist. We find the longitudes at which the rotational velocities change sign at the correct longitudes of centre and anticentre, known from other studies. The function $\omega(R)$ has been determined from the observations themselves using the tangent method. Data from Australia, not available at the time of the Symposium,* indicate a close equality between the functions $\omega(R)$ determined in the southern and the northern sky so that the motion is circular, not spiral. 'Forbidden velocities', e.g. positive velocities in the Perseus quadrant, can be ascribed to cloud motions of the order of 5–10 km/sec with occasional suggestions of local deviations of a more systematic character, up to 10 km/sec.

The reason why regions (a) and (b) have been mentioned separately is that in (a) the $r(v)$ relation is unique and in (b) it is not. Here r is the distance to the sun; the distance R to the centre is uniquely determined in both regions. Perhaps we may pass this point as a technicality although it has formed a tough problem in the reductions. As there is no indication of a structural change at $R = 8$ kpc we can discuss the results of regions (a) and (b) together.

The main results are shown in the map of Fig. 1. This is a contour map of the 'projected' densities: at each point x, y, the maximum density found on a line in the z-direction (normal to the galactic plane) is represented. As these points do not lie exactly in a plane, the map does not represent the true cross-section with any one plane. However, these points determine fairly well what might be called the true galactic plane. It is $1°5$ below Ohlsson's plane in the centre direction, $1°3$ below it near $l = 0°$, about $1°$ or $2°$ above it near $l = 80°$, about coincident with Ohlsson's plane near $l = 200°$, and again $1°$ or $2°$ below it near $l = 270°$.

* Letter from F. J. Kerr to M. Schmidt, 27 October 1955.

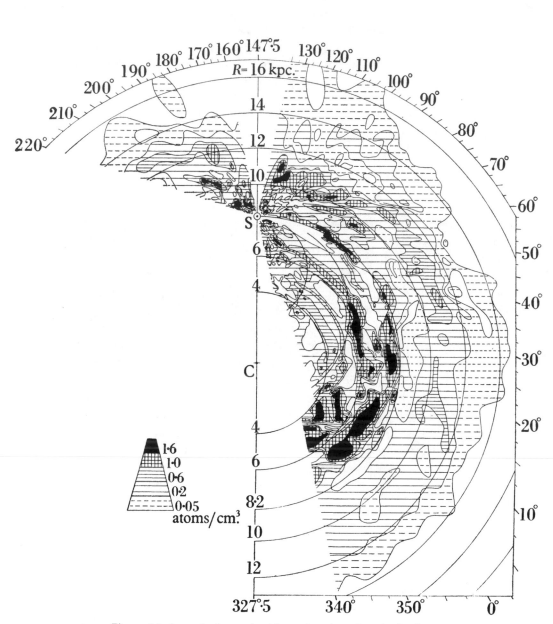

Fig. 1. Maximum hydrogen densities projected on the galactic plane.

18

The map shows the same spiral arms that were first plotted three years ago. The well-pronounced arm through the sun is the Orion arm, the heavy arm through the point $l = 100°$, $r = 2$ kpc is the Perseus arm, named after the double cluster near that point, which seems to be its most prominent association. While the coincidence is striking in this projected map, it is not on the three-dimensional model. The maximum hydrogen density in this region is reached at $b = 1°$ and the suggestion that the main arm is really there is supported by the two strong emission nebulae, IC 1805 and IC 1848 at $b = 1°5$. The double cluster is at $b = -3°5$, i.e. fully 5° away and has no emission region on the Palomar Schmidt plates nor 21-cm emission in the Leiden survey. However, local peculiarities are not excluded. This is also illustrated by a splitting of the Perseus arm from $l = 65°$ to $l = 90°$.

The arm beyond the Perseus arm is more pronounced in this map than in the published ones as it is about 200 pc above Ohlsson's plane. The maximum density in it is of the order of 0·4 atoms/cm³, i.e. roughly one-quarter of that in the Perseus and Orion arms.

Three heavy arms are seen far from the sun at longitudes between 10° and 340°. Two are inside $R = 8$ kpc and one outside. It is quite tempting to call these the extensions of Morgan's Sagittarius arm, Orion arm and Perseus arm, respectively. Perhaps this is correct, but the word spiral arm (and the theories proposed for the origin of spiral arms) is dangerously suggestive. So far, the observations of our own Galaxy do not prove that it is possible to keep track of the identity of a spiral arm after tracing it over more than 60° of galactocentric longitude. There are plenty of local bulges and deviations and for the present argument it makes little difference whether these are deviations in position or in velocity. For instance, a particularly confused region occurs near $l = 30°$ to 40° and I am not certain that it is possible to continue the Perseus arm beyond that point, although another arm takes its (approximate) position.

The densities in the arms between $R = 4$ and 8 kpc seem to be slightly higher, systematically, than in those beyond $R = 8$ kpc. This is clearly shown by the map. The cross-sections through the spiral arms resulting from Westerhout's work are, in general, elongated. The diameter of the arm in the plane is of the order of 600 pc, while the diameter normal to the plane is closer to 200 or 300 pc. These diameters are measured between points where the density is half the central density. The dimensions appear remarkably constant throughout the region of the Galaxy in which spiral arms have been observed. In this estimate, I have corrected for the slanting direction of the line of sight through the spiral arm. Also, we have tried very hard to remove the ellipticity by higher corrections for random cloud motions, but this does not seem possible. The arms seem definitely flattened. There is no theoretical reason against this.

The two other regions of the Galaxy mentioned in my introduction, (c) and (d), have in common line profiles with long wings of low intensity. To observe these wings requires a very stable receiver with a reproducible zero at each frequency. Although it must be more profitable to work with a wide band width on these wings, the present receiver was quite suitable for a reconnaissance.

The preliminary results for the central regions ($R < 3$ kpc) have been published in *B.A.N.* no. 458. The wings extend up to 150 or 200 km/sec from the zero frequency. In several cases they show up at both the positive and negative velocity side, and they are still present at the centre, where the differential velocities are 0. The tentative interpretation is that the central region of our Galaxy contains gas with high turbulent velocities. It has not yet been possible to study the state of rotation of this gas in any detail. Also the provisional continuum measurements seem to indicate a fairly sharp change near $R = 3$ kpc. The region within this distance may be virtually devoid of the sources (ionized gas) that provide the continuous radio emission in the decimetre range in the outer parts. A further study of the intriguing problems connected with the central regions of the Galaxy has been postponed until the new 25-metre telescope is available.

During the last months of measurements at Kootwijk some sample line profiles were measured at points scattered over the sky at latitudes from 30° to 90°. Half of them show central intensities corresponding to $T = 10°$–25° near the zero frequency. These peaks may be explained by relatively near clouds. However, they all show extended wings, with temperatures of the order of 2°, i.e. nearly invisible. These wings range on the average from -42 to $+24$ km/sec. The extreme velocities are roughly 50 % larger at either side. The asymmetry is quite pronounced and does not seem to be systematically related to differential galactic rotation. It seems quite likely that these wings are emitted by clouds having velocities of the order of 40 km/sec perpendicular to the galactic plane. Such clouds would reach altitudes of the order of 1000 pc and form something intermediate between the disc population and the halo population of the Galaxy. In any case, the gas emitting these wings should not be confused with the gaseous corona of the Galaxy postulated for the interpretation of the background continuum at radio frequencies. The smeared-out density of the gas emitting the 21-cm wings at high latitudes would be of the order of 0·1 atom/cm^3. The systematic velocity of approach is not well understood. The explanation by a difference in cloud concentration or in ionization between the gas travelling away from the plane and towards the plane does not seem plausible. Perhaps the systematic decrease in angular velocity at these heights may prove a better explanation, although it cannot be the only cause.

Discussion

W. W. Morgan: Professor van de Hulst's most interesting paper raises an important question of terminology: Are the galactic spiral arms defined uniquely by the 21-cm hydrogen observations? So long as there is sensible agreement in the delineation of the arms by the hydrogen gas and by the blue super-giants, this question is not of much importance; however, in the region of the Double Cluster in Perseus there seems to be a difference; it would appear, therefore, that there might be some justification for distinguishing between 'gas arms' and 'star arms'. The overall agreement between the two seems well established, but the evidence is now rather strong for some local differences.

From the Leiden 21-cm observations it appears that the Double Cluster is located in a region of low hydrogen density. However, in its general neighbourhood a number of early O-stars are observed. (These may have no physical connexion with the Double Cluster itself, but they seem also to be located in a region of low interstellar hydrogen density.) We thus have to explain the presence of a rich concentration of blue giant stars in a region of low interstellar gas density.

The radial velocity measures give strong evidence that the Double Cluster is in the general region in space of the Perseus arm as defined by the 21-cm results. The mean radial velocity for seventeen blue super-giants in the Double Cluster association and for seven red super-giants is close to the 21-cm value for the Perseus arm. In addition, the intensities and velocities of the components of the interstellar lines of Ca II and Na I by G. Munch give strong evidence that the Double Cluster is in the general neighbourhood of the Perseus arm.

This question of terminology is especially important at the present time, when, because of limitations in observational resolving power, it is necessary to compare 21-cm results for our Galaxy with optical results for the nearer spirals, like the Andromeda nebula. It would appear to be of the greatest importance to evaluate the degree of accordance of the results of the two methods for our Galaxy.

van de Hulst: There is indeed a conspicuous absence of any H I condensation in the region of the h and χ Persei cluster as indicated in the present survey. However, in general, the agreement with the H II regions, and probably also with the O-associations in these H II regions, has been very good. Would not Dr Morgan agree that the Double Cluster is a rather exceptional object with very few O-type stars?

Morgan: There are around ten early O-stars in negative galactic latitudes between longitudes 95° and 107°; it is possible, however, that none of them are directly associated with the Double Cluster. On the other hand, it is exceedingly probable that most of them are concentrated at a distance from the sun similar to that of the Double Cluster. Now, if we omit the 21-cm hydrogen, it also appears from the H II regions themselves that this is a region of low gas density because these O-stars do not have H II regions around them. My only point is that there may be extensive concentrations of early-type stars and even O-stars where there is not a concentration of hydrogen gas.

Bok: I should mention here that Dr Shajn stated very clearly a few years ago that, from his studies at the Crimea Observatory, this region was unusually free of ionized hydrogen.

Spitzer: What is the extension of the hydrogen gas perpendicular to the galactic plane?

Schmidt: The distance between points, where the density is half the density in the plane, is 220 pc.

Greenstein: The dust arms do not coincide with the B-star arms in general. This is essentially what Dr Baade pointed out.

Baade: I think that the dust arms and the gas arms in the Andromeda nebula do coincide. They give this picture: where we have H II regions, we find exactly the same arm with the same extent from these regions and from the dust itself. That is, the H II regions and the dust arms are exactly co-existent.

Greenstein: I was talking about the star arms. Even within the short span of time for the formation, life and evolution of luminous stars it is very possible that there will be appreciable separation of newly formed objects from the dust and gas regions. I believe it would be over-simplifying the situation we know to exist in our own Galaxy if we said that extra-galactic nebulae have these stars only in dusty regions.

Baade: Well, there, of course, we have the odd situation in Andromeda, that in the inner arms, we essentially see the stars only at the outer fringes because the rest is too obscured. Inside, we see very little, although we do see occasional H II regions.

de Vaucouleurs: I should like to point out in connexion with Dr Morgan's remarks, that in the Magellanic Clouds, the faint outer arms do not seem to contain strong concentrations of hydrogen gas.

Blaauw: Can anything be said about the systematic differences between the random cloud velocities in different parts of the spiral arms?

van de Hulst: This question of random cloud motions has been a great worry for us because they are needed in the reductions, or everything is blurred in the radial direction. But the data give no information except in those regions where we have what might be called forbidden velocities, in which differential rotation predicts positive velocities though we observe negative tails and conversely. By necessity these regions are quite close to the sun. In these regions there is no definite evidence of systematic differences with distance from the galactic centre. However, perhaps a theoretical argument, due to M. Schmidt, may be used. If the distance scale in the z-direction is the same at various distances from the galactic centre and if the gravitational gradient increases strongly, then the velocities must also increase in going towards the centre. This would link up with the fact, which is definitely found from the observations, that quite near the centre, very high velocities are suddenly found. So, perhaps the random cloud velocities increase from a mean of 5 km/sec, a typical value near the sun, to, say, 10 km/sec at 3 kpc from the centre, but we do not know much about it, actually.

6. SPECTROSCOPIC EVIDENCE BEARING ON THE DISTRIBUTION OF GAS IN EXTRA-GALACTIC SYSTEMS*

N. U. MAYALL

Lick Observatory, University of California, U.S.A.

Six years ago, in a joint paper with W. Baade[1], I presented certain spectrographic data bearing on the distribution and motions of gaseous masses in spiral nebulae. Since then progress in this special field has been unspectacular and relatively slow by conventional optical means, mainly because of the considerable faintness of the sources. Nevertheless, the optical data now seem sufficiently representative to warrant a review directed toward an eventual comparison of optical and radio results. For there can be little doubt that the new radio astronomy techniques give every indication of revolutionary new advances. Highly significant results have been obtained for the Magellanic Clouds, a beginning has been made in the local group spirals such as the Andromeda nebula, and it is reasonable to expect, as more powerful installations come into operation, that the nearer field and great cluster nebulae will come under detailed radio observation. But these matters have been considered at the recent Jodrell Bank symposium on radio astronomy, in Manchester. For this reason and because I am not a radio astronomer I shall proceed, without further discussion of these exciting new developments, to a description of the more prosaic optical data obtained from spectrographic observations.

When the slit of a spectrograph is placed on the optical image of an extra-galactic nebula to learn something about the distribution of gas in the system, the resulting spectrogram is examined for evidence of emission lines. These come from interstellar gas that has been ionized by the far-ultra-violet radiation from imbedded hot stars. Although the principal constituent and source of free electrons in the gas is hydrogen, other much less abundant elements reveal their presence by forbidden radiations. Among these is a close-doublet emission line normally at wave-length λ 3727, which is due to [O II]. From the spectra of more than 800 extra-galactic nebulae observed during the past twenty years at the Mount Wilson-Palomar and Lick Observatories, this particular emission feature

* *Contributions from the Lick Observatory*, Series II, no. 81.

has been found to occur with higher frequency than all others, including Hβ and Hγ, in the blue-violet spectral region [2,3]. In the red, Hα with its flanking companions of [N II] at λ 6548 and λ 6584, may rival λ 3727 as persistent emission [4], but spectrographic observations in the red are not nearly so numerous. With this limitation, λ 3727 is presently the best-known optical indicator of ionized gas in extragalactic systems.

To begin this discussion of gas in extra-galactic systems, it should be emphasized that observations of emission lines in their spectra show where the *combination* of hot stars and ionized hydrogen gas exists. There may be cases of systems in which the gas is present, but unless there are also hot stars mixed with it to effect the ionization required to produce the emissions, the optical spectra do not show directly the presence of gas. Thus the optical data give information chiefly on the distribution of Strömgren's H II regions, whereas the radio observations at 21-cm yield results on the distribution of neutral hydrogen, H I. Indirectly, the gas in extragalactic systems may be included in estimates of mass obtained from spectrographic measurements of rotational and internal motions. These optical data, however, are at present too scanty and of too low precision for a discussion of the mass ratio between stars and the interstellar matter that contains gas and dust.

One of the principal impressions obtained from detailed study of extra-galactic systems is the great diversity of structural forms. These range from smooth, featureless elliptical systems, through the regular and barred spirals of varying degrees of complexity, to irregular objects of chaotic and contorted shapes. Although almost every combination of elliptical, spiral and irregular systems can be found, Hubble has proposed a classification scheme that has been generally useful for a large majority of the brighter and nearer nebulae. These are the ones best adapted for detailed spectrographic study of emission radiations, and this discussion is mainly based on the distribution of λ 3727 radiation according to Hubble's classification. The frequencies of occurrence, from the Mount Wilson-Palomar and Lick spectrographic data [5], are shown in the table on p. 25.

This table confirms with appreciably more material a result previously obtained from fewer data, namely, that there is a definite increase in the frequency of occurrence of λ 3727 for a progression from elliptical to spiral systems. The fifth and sixth columns show as a percentage the order of the increase. They also show some systematic differences between the two sets of data, but these differences are not so large as to invalidate the trend, and they may reasonably be ascribed to the different circumstances of observation [1]. If averages are taken, the frequency increase is from about 15 % for ellipticals to nearly 80 % for spirals, or by a factor of approximately 5.

The foregoing overall statistics, however, give little hint as to how the ionized gas, as indicated by the presence of λ 3727, may be distributed in the various types of extra-galactic systems. The Lick spectrographic material has been examined for this purpose, with details given in a table of red-shifts [5]. For the present discussion, it seemed most practicable to present typical results in a series of illustrations, first individually described and then collectively considered as to some possible general conclusions. The procedure is according to Hubble's nebular types, beginning with ellipticals, followed by normal spirals, barred spirals, apparently close doubles, and finally two streamer-connected pairs.

Frequency of occurrence of λ 3727

Type	Lick				Mount Wilson-Palomar	
	λ 3727/all	(%)	λ 3727/total	(%)	(%)	λ 3727/all
E	5/41	12	5/41	12	18	15/82
So	6/23	26⎫				
SBo	3/10	30⎭	9/33	27	48	25/52
Sa	7/16	44⎫				
SBa	2/4	50⎭	9/20	45	62	23/37
Sb	31/46	67⎫				
SBb	9/19	47⎭	40/65	62	80	53/66
Sc	73/106	69⎫				
SBc	8/12	67⎭	81/118	68	85	35/41
Irr	16/17	94⎫				
Spec, Pec⎱	4/6	67⎭	20/23	87	—	—
Sab, Sbc⎰						
All	—	—	164/300	55	55	151/278

Fig. 1 shows in the upper panel spectra of two ellipticals, in one of which, NGC 5846, λ 3727 may be faintly present, because measurement of the feature marked with an arrow gave a red-shift in agreement with those for the H- and K-absorption lines. The spectrum of the uncatalogued companion above 5846 shows a similar emission feature (of a little shorter wave-length) that is not λ 3727 because it does not give a red-shift in agreement with a larger red-shift measured for the H- and K-lines. In both cases, however, these features that appear as faint emission in the general region of λ 3727 could be only intensifications in the continuous spectra, and further observations are needed to reach a decision. In the central panel, λ 3727 is clearly present in the isolated elliptical NGC 1453; the smaller direct image is due to a too-weak exposure. The lower panel shows the non-typical but interesting case of NGC 4486 (M 87), which has been discussed as a radio source by Baade and Minkowski[6]. Here the slit was on the nucleus and along the ray or jet, which may be seen projecting downwards in the direct photograph. The spectrogram shows λ 3727 fairly strong in the nucleus, and probably absent in the ray. The emission lines extending entirely across the spectrum are 4047 and 4358 of Hg I, which come from the low, western horizon sky-glow produced by city lights in the Santa Clara valley.

Fig. 2 shows how λ 3727 appears in three early-type spirals, two classified by Hubble as So and one as Sa. In all three cases λ 3727 is not only present in faint (NGC 3169) to strong (NGC 3998) intensity, but it likewise shows rotational inclination comparable to the principal absorption lines. These spectra also show the tendency, only suggested by the ellipticals, for the λ 3727 radiation to be concentrated toward the center of the systems. For all three of these systems the spectrographic exposures were long enough to have shown λ 3737 well beyond the nuclei, if it were present.

Fig. 3 shows three intermediate-type normal spirals, all Sb in Hubble's classification scheme. One of these, NGC 2841, has a prominent nuclear bulge, while the other two, with less extensive central parts, instead have well-developed, inner spiral-ring structure. In the upper panel, λ 3727 appears in moderate intensity only in the nuclear region, as was the case in the elliptical and earlier-type spirals considered in Figs. 1 and 2. On the original spectrogram of NGC 2841 the night-sky spectrum shows strongly enough to suggest that, if λ 3727 were present in appreciable intensity in the outer parts of this spiral, the exposure was long enough to have recorded it. The central and lower panels are intended to show that λ 3727 may sometimes be stronger in an interior spiral-arc ring, as in NGC 4736 (M 94), or sometimes in the nucleus, as in NGC 4750. While λ 3727 is also present in the nuclear region of M 94, its intensity there appears to be less than in the part of the ring that was on the slit. In all three spirals emission and absorption lines are measurably inclined because of rotation of the systems.

Figs. 4 and 5 show six examples of Sc or late-type normal spirals, which, as is well known, have a great variety of structural features. These systems were selected from more than 100 observed with the Crossley nebular spectrograph in order to illustrate the considerable variation in the distribution pattern of λ 3727.

In the upper panel of Fig. 4, this emission is fairly strong in the short, bar-like nucleus of NGC 925 and in one knot, but faint elsewhere. In the central panel, λ 3727 is present in good strength throughout NGC 2748, possibly because the principal plane of the system is tilted not far out of the line of sight and the light path is longer in the spiral. For the nearly square-on spiral NGC 2835 in the lower panel, λ 3727 is present in a pronounced patchy distribution that corresponds closely with the light

26

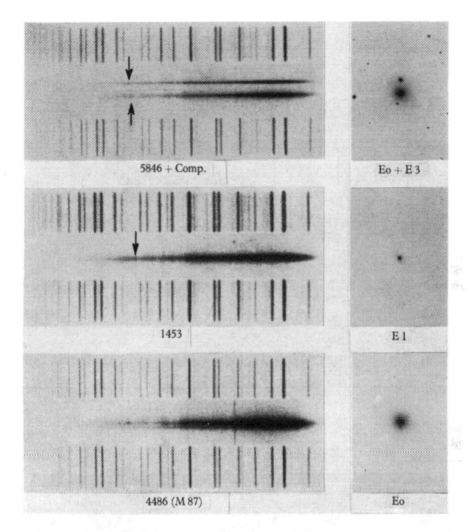

5846 + Comp.

Eo + E 3

1453

E 1

4486 (M 87)

Eo

Fig. 1.

facing p. 26

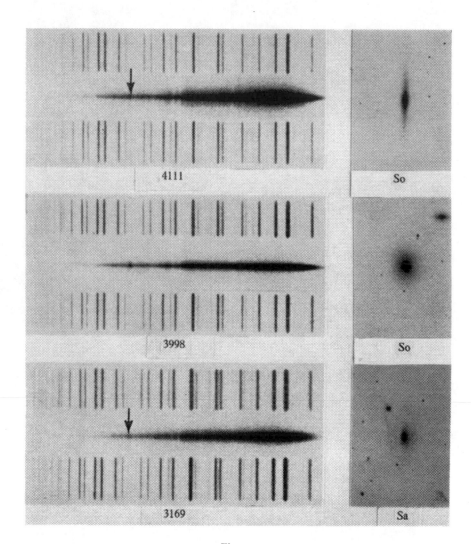

Fig. 2.

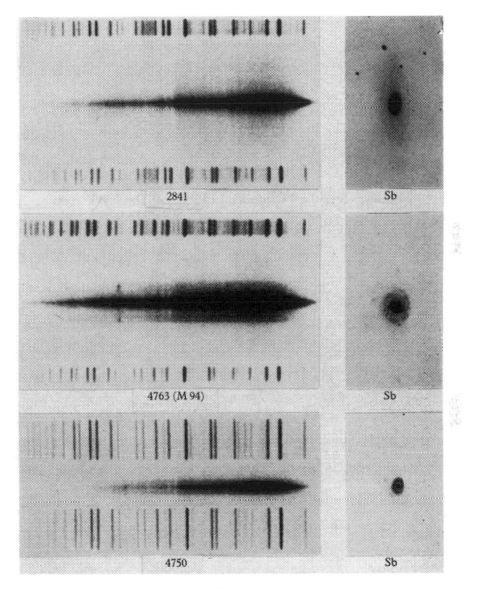

2841

Sb

4763 (M 94)

Sb

4750

Sb

Fig. 3.

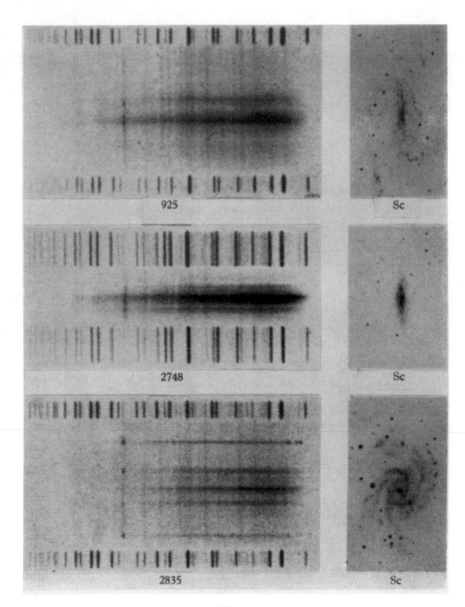

925   Sc

2748   Sc

2835   Sc

Fig. 4.

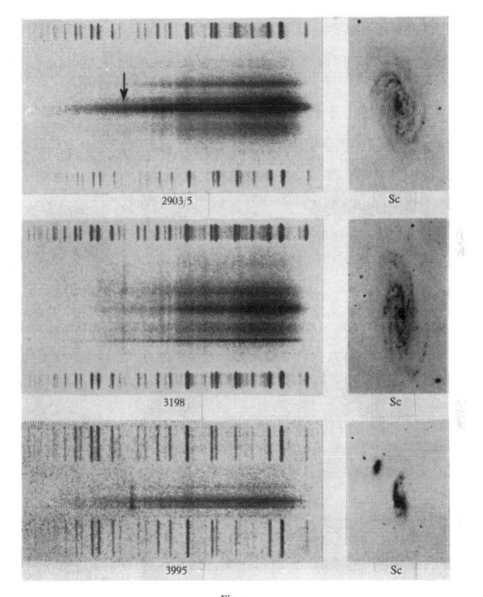

2903/5   Sc

3198   Sc

3995   Sc

Fig. 5.

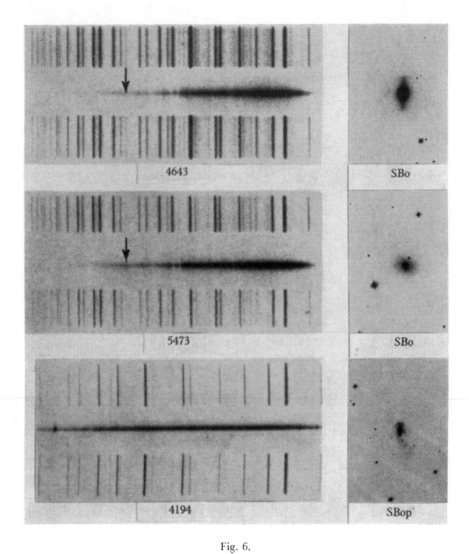

Fig. 6.

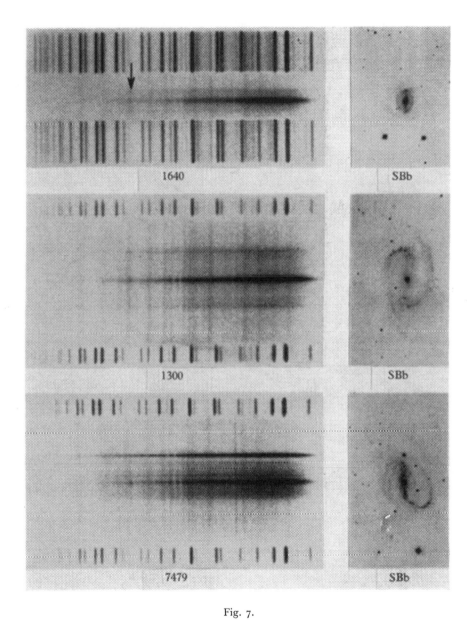

1640

SBb

1300

SBb

7479

SBb

Fig. 7.

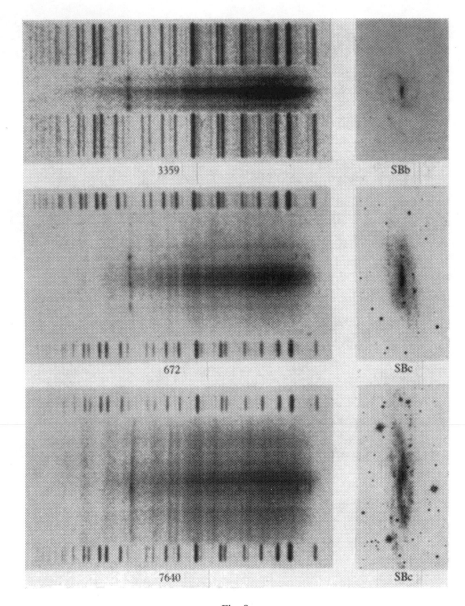

3359   SBb

672   SBc

7640   SBc

Fig. 8.

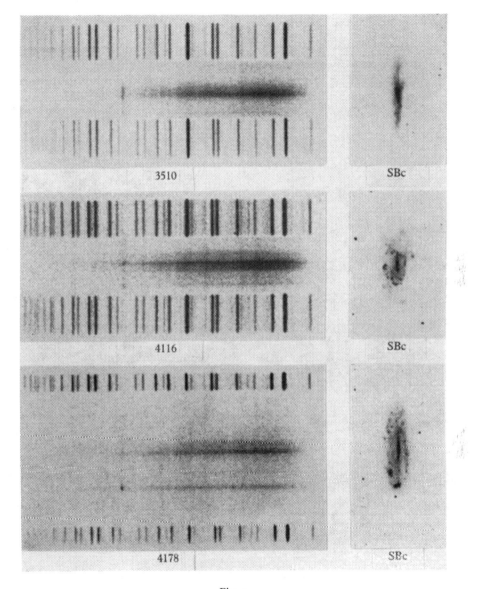

Fig. 9.

Facing Fig. 8

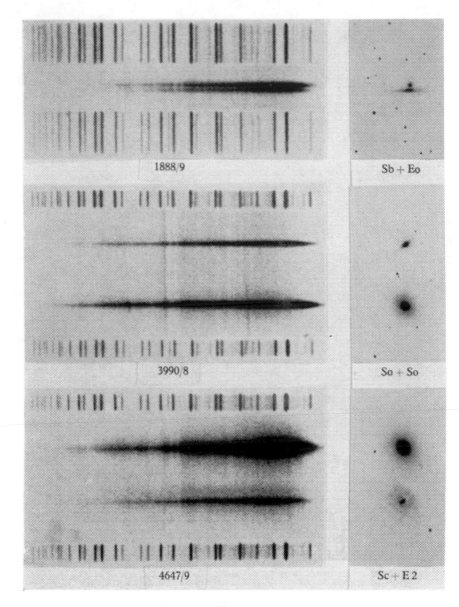

1888/9   Sb + Eo

3990/8   So + So

4647/9   Sc + E 2

Fig. 10.

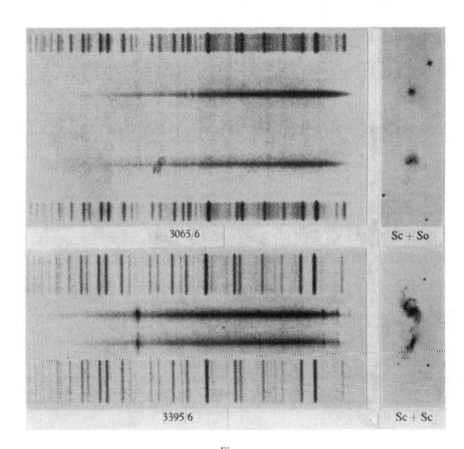

3065/6

Sc + So

3395/6

Sc + Sc

Fig. 11.

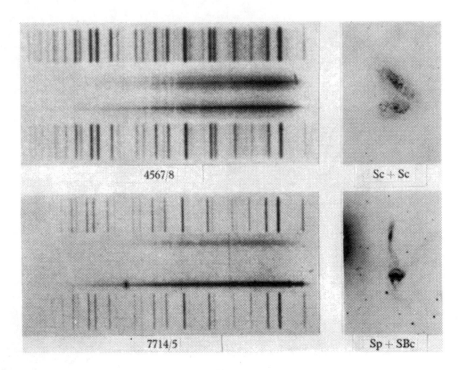

4567/8   Sc + Sc

7714/5   Sp + SBc

Fig. 12.

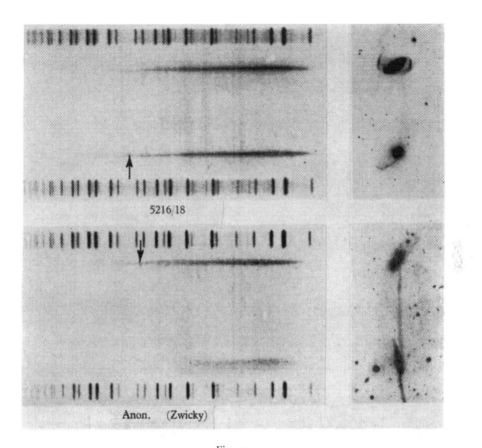

5216/18

Anon.   (Zwicky)

Fig. 13

Facing Fig 13

distribution in the direct exposure, except in the nucleus where it is much fainter or absent. For the brightest knot that was on the slit, the spectrum shows a number of other emission radiations that include those at λ 4957 and λ 5006 of [O III], with intensities relative to Hβ typical of galactic gaseous diffuse nebulae.

Fig. 5 shows examples of weakest and strongest appearances of λ 3727. In the upper panel this emission appears faintly present mainly in the nuclear region of NGC 2903/5, although a slightly different orientation of the slit might show it to occur with fairly uniform intensity throughout, as for NGC 3198 in the central panel; in the latter, the strong, narrow continuous spectrum at the lower end of the slit is that of a foreground star. In the bottom panel, λ 3727 shows in high intensity in the strongest arm and brightest condensation in NGC 3995, which appears to be the distorted member of a double system.

At this point we leave the so-called normal spirals and consider the barred spirals. Although the latter seem to have a characteristically different structure in direct photographs, spectroscopically they do not give much reason to think that they are systematically different. As indicated in the next group of illustrations, the same emission and absorption lines occur, with the same diversity in intensity, and with comparable complex relationship to structure, for the barred as for the normal spirals.

The upper and central panels of Fig. 6 show two earliest-type barred spirals, NGC 4643 and 5473, classified by Hubble as SBo. In these nebulae the very faint features marked with arrows were measured as λ 3727 and gave red-shifts corresponding to the absorption H- and K-lines. The result is thus the same as for ellipticals: if λ 3727 is present, it tends to be faint and concentrated toward the center. The case is quite different, however, for the nebula in the bottom panel. This shows a 3-prism spectrogram of the 'peculiar' SBo system NGC 4194, in which λ 3727 is not only strong, but hydrogen lines of abnormal intensity and width are present. Here Hβ and Hγ are in emission, the higher members of the Balmer series are in absorption, and the [O III] pair can be seen. There is little doubt in this case that the qualification 'peculiar' is justified.

In the Crossley collection there are at present no spectrograms showing λ 3727 in SBa systems, not because none are known, but only because no systematic program for their observation has been undertaken. Therefore, we next consider the intermediate-type barred spirals—those denoted SBb in Hubble's classification.

Fig. 7 shows three SBb's that represent the range, for this class of system, of the degree to which the spiral arms seem to have uncoiled from a ring structure that often is present around the nucleus. With the possible exception of one very faint appearance in a knot in NGC 1640, λ 3727 is rather strongly concentrated toward the nuclei of these three spirals. While other slit positions in their outer parts might show λ 3727, these regions are so much fainter in these spirals that the nebular spectral features are likely to be confused with the night sky spectrum. There is better evidence, from other cases, that λ 3727 shows no general tendency to be confined to the nuclear regions of SBb and SBc systems; some examples are given in the next two illustrations.

Fig. 8 shows, in the upper and central panels respectively, intermediate-type and late-type barred spirals, NGC 3359 and 672, in which λ 3727 is strong just beyond the ends of the bar, localized in knots or condensations, in addition to being present in the bar. For NGC 3359 the direct photo-graph is a too-weak exposure that fails to show details of the outer spiral structure where the slit crossed it. Thus the spectrogram seems to show strong emission where there is little light recorded in the direct image. In the lower panel the nearly edge-on late-type barred spiral NGC 7640 shows λ 3727 present along the bar and major axis in fairly uniform intensity that corresponds closely with the distribution of light in the direct photograph.

Fig. 9 shows three more late-type barred spirals of which two, NGC 3510 and 4116, are irregular in appearance. In these two systems λ 3727 is distributed in patchy intensity both in the bar and in the outer irregular parts. The bottom panel reproduces a grainy spectrogram (pre-War 2) that shows λ 3727 in only one bright knot near one end of the system. This is a case where a better-quality plate, with IIaO emulsion and a narrower slit, is needed to determine the presence or absence of λ 3727 in the bar. From the other foregoing examples a positive result would be expected.

We next consider the distribution of λ 3727 in some double systems in which the components are of nearly the same type, of different type, or streamer-connected. As was the case for the single systems illustrated, the selection of doubles is but a small sample from the large number of known close pairs. The variety is so great that the following group of illustrations can only suggest the richness of the field.

Fig. 10 shows three pairs of systems in which one component in each appears to be an exception to the statistical result for λ 3727 to be stronger with advancing nebular type. In the top panel NGC 1888, Sb, seems to show little or none of this emission, in the center NGC 3998, the larger and brighter So, shows λ 3727 strongly, and in the lower NGC 4647, Sc, shows the radiation only faintly. These three exceptional cases are not intended, however, to show that double systems differ systematically from single ones. The purpose is to suggest that the relationship between structure and distribution of ionized gas probably is as complex for double as for single systems.

Fig. 11 illustrates two pairs of systems that show nearly the extremes of low and high intensities of λ 3727. In the upper panel, the components are not known from direct photographs to be connected by a streamer, but the appearance of the original spectrogram in the region of λ 3727 suggests a check of this possibility. For the spirals in the lower panel, T. L. Page[4] has reported λ 3727 connecting them. This result probably would have been obtained with the Crossley nebular spectrograph if the slit were placed slightly differently and if the exposure were longer.

Fig. 12 shows two close doubles for which the spectrographic data have unusual interest. In the upper panel, there is no certain evidence of emission radiation in either component, but in the lower panel the brighter member shows exceptionally strong λ 3727 and other bright lines. This spiral is classified as peculiar because of its distorted form. Thus the very different spectral features in these two pairs may indicate the effects on the distribution of ionized gas of little and much interaction.

Fig. 13 shows the result of attempts to find, in two streamer-connected systems[7,8], spectroscopic evidence for an ionized gaseous constituent in the streamers. Examination of the original negatives, and experience with many others similarly exposed long enough to reach the background night-sky spectrum, leads to the conclusion that no emission lines are present in the blue and near ultra-violet spectra of these two pairs. Although this is a result for only two cases, the connecting filaments are among the brightest known.

In making generalizations from the spectrographic data reproduced as Figs. 1–13, it should be emphasized that any one spectrogram showing emission gives information regarding the distribution of ionized gas only in a single plane corresponding to the orientation of the slit. This, in many cases, was placed on the more conspicuous components such as nuclei, spiral arms, bars, knots, or doubles for red-shift or differential velocity measurements. These data therefore involve a great deal of observational selection, the effect of which is difficult to evaluate. A much better approach to the problem would be to use narrow-band color-filter photography, which would include a whole system. A comprehensive survey, carried out with moderate to large reflectors utilizing modern interference filters and fast plates, would place on a much firmer basis our knowledge of the distribution of H II regions in the nearer extragalactic systems of all classes. In the meantime, the spectroscopic evidence suggests the following summary conclusions regarding the distribution of ionized gas in extragalactic systems, as inferred from the occurrence of λ 3727:

(1) In the progression of nebular types from ellipticals to spirals, the gas tends to become less centrally concentrated and to occur more frequently by a factor of 5 or 6.

(2) While the H II regions are often related to such structural features as nuclei, spiral arms, bars or rings, the relationship is not very close; there appears to be little systematic difference between normal and barred spirals.

(3) For close doubles, there is some indication of increased ionization when the components either are of very late or irregular nebular type, or when peculiar structure suggests strong interaction.

(4) In two cases of wider pairs having relatively bright connecting streamers, there is no evidence for an ionized gaseous constituent in the joining filaments.

REFERENCES

[1] Baade, W. and Mayall, N. U. *Proceedings of the Symposium on the Motion of Gaseous Masses of Cosmical Dimensions, Paris, Aug. 16–19, 1949*, Central Air Documents Office, Dayton, Ohio, 1951.
[2] Mayall, N. U. *Lick Obs. Bull.* **19**, 33 (no. 497) 1939.
[3] Humason, M. L. *Publ. A.S.P.* **59**, 180, 1947.
[4] Page, T. L. *Ap. J.* **116**, 63, 1952.
[5] Humason, M. L., Mayall, N. U. and Sandage, A. R. *A. J.* **61**, 97, 1956.
[6] Baade, W. and Minkowski, R. *Ap. J.* **119**, 215, 1954.
[7] Keenan, P. C. *Ap. J.* **81**, 355, 1935.
[8] Zwicky, F. Private communication.

Discussion

Page: It is worth mentioning similar observations of double galaxies made with a small grating spectrograph at the McDonald Observatory. Many of these same objects were observed in the red region, including Hα and the forbidden N II doublet 6548 and 6584 Å. These plates showed many of the same aspects that Dr Mayall has described. As might be expected, among about 100 spectra, Hα appeared in every case where 3727 appeared, with one exception. However, in two cases, the bridge between double galaxies showed emission both in 3727 (which generally seems more extended than Hα) and in the forbidden nitrogen lines, which are similarly extended, but *not* in Hα.

Oort: Can Dr Mayall give more details about the rotation periods of the barred spirals he has measured?

Mayall: Only ten nebulae of type SBc were measured. The measurable inclinations of the spectral lines give rotation periods that vary from about 50 to

250 million years, for inclinations ranging from $1°$ down to about $0.°2$ or $0.°3$. However, there are some smaller inclinations of the lines which could not be measured. I assumed 200 km/sec/1,000,000 pc for Hubble's red-shift parameter in order to obtain distances to compute periods.

Oort: Does Dr Mayall mean that there is a big selection effect and that he could not measure the period of rotation in many cases?

Mayall: There is a big selection effect in the sense that only those nebulae were observed in which the bar was nearly normal to the line of sight and for which the spiral was favourably tilted. However, under these conditions, we get a measurable but small inclination whenever the spectrogram is well enough exposed. Since only ten late-type barred spirals have been observed, it would be very hasty to draw any generalization on the basis of our present data.

7. ON THE ROLE OF DIFFUSE EMISSION NEBULAE IN GALAXIES

THE LATE G. A. SHAJN

Simeis Astrophysical Observatory, Crimea, U.S.S.R.

(Read by Mrs Masevich)

The emission nebulae, as more or less compact gaseous or gas-and-dust formations, belong to a rather widespread type of population in our Galaxy. Though, for well-known reasons, we are able to penetrate into the depth of the Galaxy no further than 2–3 kpc for H II regions, several hundred nebulae have already been recorded (predominantly in the spiral arms). Sometimes the emission nebulae almost entirely occupy large regions of the sky, as, for instance, in the region of ρ, γ and o^1 Cygni. In addition, in the area within latitudes $\pm 5°$, we observe a nearly continuous faint emission background[1].

Emission nebulae are undoubtedly widespread in extra-galactic spiral systems also. We were able to investigate in detail with our 64-cm camera (F 1·4) only the nearest spiral systems: M 31, M 33, M 51, M 101, NGC 2403, NGC 3627, NGC 4258, NGC 4631. According to Baade and Mayall, there are hundreds of emission nebulae in M 31 located along the spiral arms[2]. Over a hundred emission nebulae were found in M 33 with the aid of moderate-size instruments[3,4,5]. By using our photographs taken with a red Hα filter, a narrow violet one [λ 3727 (O II)] and an ordinary yellow one, we were able to obtain the distribution of the emission nebulae in M 33[5]. In addition to these bright nebulae, we suggested the presence of a considerable number of fainter nebulae, and also of many patches of an emission character. In the other spiral galaxies mentioned above, we also found a number of bright emission nebulae. The majority of these were known or had been suggested earlier[6,7,8]. The faintest nebulae recognized with certainty by us are about $15^{m}·7$ (red magnitude). This corresponds approximately to the absolute magnitude $-6^{m}·0$ for M 33 and M 31 and to $-7^{m}·7$ for the other galaxies mentioned above (or $1^{m}·5$ brighter according to Baade's new distance scale). The integral absolute magnitude of the Orion nebula, NGC 1976, is about $-6^{m}·5$ at the same wave-length. Therefore, the faintest emission nebulae recorded by us are no fainter intrinsically than NGC 1976; the majority of them are brighter or much brighter than the brightest nebulae in the accessible part of the sky.

The above-mentioned spiral galaxies were selected only because they are the nearest to us. Therefore, there is no reason to consider them exceptional, particularly in respect to their gaseous population. There is no doubt that in each of these galaxies, in addition to the brighter emission nebulae which we have recorded, there are hundreds of other nebulae as in M 33 and M 31. The galaxies under consideration are all of the Sc and Sb types. It seems that M 31 is no poorer in emission nebulae than our Galaxy, while other galaxies are no poorer than M 31 and may be even richer. For example, the emission nebulae in M 33 are brighter than those in M 31 (by $1^m \cdot 0$ or $1^m \cdot 5$ in the mean). Therefore, the nearest galaxies selected for a more detailed study are at least as rich in their gaseous population as is our Galaxy; galaxies like M 33 and M 101 are much richer.

On the other hand, there are reasons to believe that the spiral galaxies under discussion do not differ at all from other spiral galaxies. This holds true particularly for the Sc galaxies. One can hardly doubt that emission nebulae are present in practically all galaxies. If these galaxies could be studied in detail, we should find them nearly as rich in nebulae as M 31, M 33, M 51, M 101, NGC 2403, etc. For this reason, one can assume that richness in nebulae distributed along the spiral arms is a characteristic feature of Sb and Sc galaxies. The problem of emission in the central part is not so clear.

The peculiarities of the spiral galaxies discussed above are of great interest in connexion with the question of the relative age of the galaxies as a whole and of the emission nebulae contained in them. Usually 3×10^9 years is assumed as a lower limit for the age of galaxies in the stage we are now observing. On the other hand, there are important reasons to believe that the age of the diffuse emission nebulae, as more or less compact formations, does not exceed 10^7 years, i.e. that it is about a hundred times less than the age of the galaxy as a whole.

Some arguments in favour of the short age of emission nebulae are given below.

(*a*) The observed structure of several emission nebulae in the form of rings or arcs may be interpreted as an indication of the outward motion of the matter and, therefore, of a recent origin of these nebulae as well as of their dissipation in the near future [9].

(*b*) The existence of groups of emission nebulae for which there are serious reasons for believing that the total energy is positive [9].

(*c*) The expected expansion of the emission nebulae and their dissipation as a result of temperature and pressure differences between the nebulae and the surrounding cold regions. Even the strong magnetic field ($\sim 10^{-4}$ gauss) in the inner part of the Orion nebula does not prevent it from expansion and dissipation [9].

(*d*) The disintegration of emission nebulae because of collisions (once in 10^7 years on the average). This argument was suggested by Oort for interstellar clouds in general [10].

The result depends to a large extent on whether or not the nebula and its exciting star are connected genetically. There are important reasons to think that the connexion is genetic, but even if it is only temporary and accidental, the nebula has enough time to expand and dissipate before getting out of the ionization zone of the exciting star. It is easy to see that doubling the size of a nebula leads to a decrease in the emission measure by a much greater factor (thirty-two times).

These arguments lead to the conclusion that the age of the diffuse nebulae in the emission stage is really short, of the order of 10^7 years or less, i.e. about a hundred times less than the age of the galaxies as a whole.

This result must be considered reliable. It is also necessary to consider as certain, the result that all spiral galaxies, or at least the overwhelming majority of them, are very rich in emission nebulae and that the latter are located along the spiral arms to a great extent. There are two possible explanations:

(1) We observe our Galaxy in a peculiar stage when these short-lived objects are formed. But emission nebulae occur in all spiral galaxies. This means that if this hypothesis is true it is necessary to assume that such a peculiar stage has set in for all galaxies at the same time. Such a supposition seems quite improbable.

(2) Emission nebulae originate more or less continuously, after which they disintegrate, dissipate, or are possibly transformed into other objects. There are some reasons to conclude that nebulae originate in groups. Some important consequences of this hypothesis must be emphasized now. From the age of the diffuse nebulae ($\sim 10^7$ years or less) and their mean spatial velocity of 10 km/sec or more, we find that the nebulae have been formed not far from their present locations in the spiral arms, probably within several hundred parsecs. But we observe nebulae along the entire spiral arms. Therefore, we must assume that the nebulae have originated and continue to originate almost everywhere along the spiral arms. Evidently, we must also conclude that the spiral arms themselves are

very steady formations, notwithstanding the short age of their constituents (gas, dust, and hot stars) [9].

If the age of emission nebulae is shorter than the age of the galaxies by at least a hundred times and the nebulae themselves disintegrate and dissipate within this period (less than 10^7 years) it is to be expected that a great quantity of matter, at least a hundred times as much as is now present in the form of emission nebulae, has dissipated during the age of the Galaxy ($\sim 3 \times 10^9$ years). We cannot estimate the total mass of nebulae, but if it is of the order of a few per cent of the total mass of the stars, the total mass of the material scattered during 3×10^9 years would be more than the total mass of all the stars.

If the dissipation is a one-sided process, we should arrive at a contradiction with the well-known upper limit for the mass of diffuse matter derived by Oort.

At present we can observe only the process of disintegration and dissipation of the emission nebulae, but it seems probable that some other kind of transformation of the diffuse matter in the form of ions, neutral atoms and dust also exists, a process which must be intimately related to the formation of stars.

In the problem under consideration, the possible effect of the interstellar magnetic field deserves attention. It is known that the magnetic lines of force seem to be oriented along the spiral arms. In addition, we found recently that the magnetic field is probably responsible for the greatly elongated shape of many emission and dark nebulae. Therefore, it may be sufficiently effective to control the distribution and motion of diffuse matter in the Galaxy to some extent [11]. If so, the magnetic field will probably contribute to the conservation of the diffuse matter as a system and prevent it from scattering freely, notwithstanding the above-mentioned dissipation of the emission nebulae. In any case, the scattering will probably be restricted in the presence of a sufficiently strong interstellar magnetic field. This means that the magnetic field will contribute to the conservation of the spiral structure of galaxies, at least their gas and dust population. However, the emission nebulae and the hot stars are so intimately related that the last suggestion probably applies to the O–B stars also—another important constituent of the population of the spiral arms.

REFERENCES

[1] Shajn, G. A. and Hase, V. F. *Atlas of Diffuse Gaseous Nebulae*, 1952.
[2] Baade, W. and Mayall, N. U. *Probl. of Cosm. Aerodyn.* 1951, p. 165.
[3] Haro, G. *A.J.* **55**, 66, 1950.

[4] Aller, L. H. Manuscript, Michigan, 1951.

[5] Shajn, G. A. *Comptes rendus de l'Acad. d. Sci. de l'U.R.S.S.* **93**, no. 6, 993, 1953.

[6] Mayall, N. U. and Aller, L. H. *Publ. A.S.P.* **51**, 112, 1939.

[7] Seyfert, C. W. *Ap. J.* **91**, 528, 1940.

[8] Courtès, G. *C. R.* **234**, 506, 1952.

[9] Shajn, G. A. and Hase, V. F. *Comptes rendus de l'Acad. d. Sci. de l'U.R.S.S.* **96**, no. 5, 945, 1954; no. 6, 1129, 1954; *Publ. Crim. Obs.* **8**, 80, 1952.

[10] Oort, J. H. *Probl. of Cosm. Aerodyn.* 1951.

[11] Shajn, G. A. *A.J. U.S.S.R.* **32**, 110, 1955.

8. GALACTIC AND EXTRA-GALACTIC RADIO FREQUENCY RADIATION DUE TO SOURCES OTHER THAN THE THERMAL AND 21-CM EMISSION OF THE INTERSTELLAR GAS

R. HANBURY BROWN

Jodrell Bank Experimental Station, University of Manchester, England

I. INTRODUCTION

At wave-lengths greater than about one metre the majority of the radio emission which is observed from the Galaxy cannot be explained in terms of thermal emission from ionized interstellar gas. This conclusion is widely accepted and is based on observations of the equivalent temperature of the sky and the spectrum of the radiation. The spectrum at metre wave-lengths is of the general form:

$$T_A \propto \lambda^n,$$

where T_A is the equivalent black-body temperature of a region of sky and λ is the wave-length. The exponent n varies with direction but lies between about 2·5 and 2·8, and is thus significantly greater than the value of 2·0 which is the maximum to be expected for thermal emission from an ionized gas. Furthermore the value of T_A is about $10^{5°}$ K at 15 m and thus greatly exceeds the electron temperature expected in H II regions.

At centimetre wave-lengths it is likely that the majority of the radiation observed originates in thermal emission from ionized gas; however, the present discussion is limited to a range of wave-lengths from about 1 m to 10 m where the ionized gas in the Galaxy is believed to be substantially transparent and where the origin of most of the radiation is believed to be non-thermal.

2. SOME FEATURES OF THE GALAXY AT METRE WAVE-LENGTHS

(a) The general background radiation

Early surveys of the sky showed that the general radiation is, broadly speaking, concentrated in latitude about the galactic plane and in longitude about the galactic centre. On the basis of these surveys, it was concluded that the sources of emission in the Galaxy, whatever they may be,

37

have a space distribution like that of the common stars. Irregularities in the distribution, which are increasingly pronounced as the wave-length is decreased, were attributed to the effects of spiral structure in the Galaxy.

The early surveys were made with rather wide beams and it is now known that much important detail was lost. Thus a recent high resolution survey by Scheuer and Ryle, and also some unpublished work by Mills, suggest that the true distribution of intensity normal to the galactic plane is made up of perhaps three component distributions. Two of these components are narrow and have widths of about 2° and 10°. It is difficult, without more extensive surveys, to be sure of their independent existence and this is a problem which requires a considerable amount of further study. The third component appears to be much broader and to have a width of the order of 60°. The variation of these components with galactic longitude is not yet known satisfactorily and is clearly complicated by irregularities in galactic structure; nevertheless it is known that the narrow distributions show a marked concentration towards the galactic centre, while the broader component appears to be concentrated in that direction to a lesser extent.

The space distribution of the sources of the broad component presents a particularly interesting question, and it is difficult to escape the conclusion that the Galaxy has a radio corona which extends to great distances. In the earlier interpretations of the isophotes it was found that agreement with the general distribution of mass in the Galaxy could be obtained by assuming that a large fraction, about two-thirds, of the total background radiation was isotropic and probably of extra-galactic origin. It seems likely that a substantial fraction of this isotropic component must now be attributed to the Galaxy and associated with the broad distribution.

(b) The discrete sources at metre wave-lengths

The first surveys of the discrete sources or radio stars were made with interferometers and it was concluded from these results that the distribution of sources with direction from the sun is isotropic. While this is still believed to be true of the majority of sources, it is now recognized that the resolving power of the interferometers was so high that important sources of large angular diameter were missed. It is now apparent that there are at least two classes of source.

Class I sources, which form a minority of the total, are of relatively high intensity and show a pronounced concentration into the galactic plane. A surprising feature of these sources is that many of them are known to have apparent angular diameters greater than one degree.

Class II sources, which form the majority, appear to be uniformly distributed over the sky. The angular diameter of many of these sources is not yet known and it is important that they should be measured; however, the few results which are available suggest that, for the most part, their diameters are of the order of a few minutes of arc or less.

3. THE ORIGIN OF THE RADIO EMISSION FROM THE GALAXY AT METRE WAVE-LENGTHS

Any discussion of the origin of the radiation from the Galaxy must be highly speculative, since recent work has shown that our knowledge of the distribution of the background radiation is seriously incomplete. Furthermore, data on the spectra and angular diameters of the sources are confined to a few of the most intense.

The present evidence suggests that any theory may have to account for both the broad and narrow distributions, although it is by no means clear whether these distributions can be regarded as independent.

Two components of the narrow distributions are the thermal radiation from ionized gas in H II regions and class I sources. At metre wave-lengths the thermal radiation cannot account for the total intensity observed and it is tempting to ascribe the remainder to a population of sources which lie close to the galactic plane. However, this cannot be done, since so little is known about the class I sources; for example, it is not known whether they are a homogeneous population, nor how they are distributed in the Galaxy. Until more data are available we must be prepared to find that some other mechanism, as yet unknown, is responsible for the majority of the radiation. For example, it was suggested by Ginsburg some years ago that the non-thermal radiation might be due to cosmic ray electrons in interstellar magnetic fields.

The nature of the known class I sources is a fascinating problem. They appear to be rare bodies with a space-density in the neighbourhood of the sun which we may compare, solely for the purpose of illustration, with that of planetary nebulae. The spectra and apparent surface temperatures of a few of these sources is known and it is clear that some of them are radiating by a non-thermal mechanism. The large angular diameters of several of these sources, coupled with the few photographic identifications which have been made, suggest that they are associated with extended nebulosities. These nebulosities are of low photographic brightness and some of them have been found to contain filaments which are apparently moving at very high speeds. The nature of these nebulosities is controversial. At least two, if not three, of the sources have been identified with the remnants of super-novae, and it has been suggested that many of the other sources arise in the same way.

The physical mechanism by which these nebulosities radiate is also unknown. It seems likely that plasma oscillations cannot be invoked since the plasma frequency in the medium is too low, and the current idea is that the radiation arises from the deflexion of relativistic electrons in magnetic fields. The magnetic fields are presumed to be generated by turbulence in an ionized medium, and the fast electrons to be accelerated by the Fermi mechanism, by shock waves, or by some other process.

The origin of the broad distribution is also a challenging problem and it is an urgent task of observation to establish beyond doubt the shape and spectrum of this distribution. It has been suggested that, whereas the narrow distributions are apparently associated with populations concentrated into the galactic plane, the broad distribution arises in an extended halo which is roughly spherical and extends to radial distances of the order of 10,000 pc. It has also been proposed that the generation of the energy in this halo occurs in a very rarefied medium and is due to the deflexion of fast electrons in magnetic fields.

The origin of the majority of the discrete sources, the class II sources, may be extra-galactic. Recent work has shown that their distribution is remarkably isotropic and it is difficult to associate them with any of those components of the background radiation which are clearly of galactic origin.

4. EXTRA-GALACTIC SOURCES

A small number of radio sources have been identified with external galaxies, and on the basis of these results it appears that, as far as radio emission is concerned, we must recognize at least two major classes of galaxies, *normal* galaxies and *peculiar* galaxies.

(a) Normal galaxies

If the apparent radio magnitude of a galaxy is designated by m_R and its photographic magnitude by m_p, then it has been found for Sb galaxies that at a wave-length of about 1·9 m these two quantities are approximately equal if m_R is defined by the equation:

$$m_R = -53\cdot4 - 2\cdot5 \log I,\qquad (1)$$

where I is the intensity in watts m^{-2} $(c/s)^{-1}$.

Table 1 shows six type Sb galaxies which have been reasonably well identified.

The values of m_R shown in the table have been calculated by means of equation (1), while the values of m_p have been taken from de Vaucouleur's revision of the Shapley-Ames catalogue. The data for NGC 1068 and I 5267 have been taken from some unpublished work by Mills. The value for NGC 224 is probably too great, due to a failure to integrate satisfactorily over the whole source, and it is likely that the true value of $m_R - m_p$ for this galaxy is closer to $+1\cdot0$.

Table 1. *Sb galaxies*

Galaxy	Radio magnitude m_R ($1\cdot9$ m)	Photographic magnitude m_p	$m_R - m_p$
NGC 224 (M 31)	6·0	4·0	+2·0
NGC 3031 (M 81)	8·9	7·8	+1·1
NGC 4258	9·8	9·1	+0·7
NGC 2841	10·4	10·2	+0·2
NGC 1068	8·9	9·6	−0·7
I 5267	11·1	10·8	+0·3

The data presented in the table, although admittedly scanty, indicate that type Sb galaxies radiate roughly the same ratio of radio to light, and that equation (1) may be used to calculate their approximate radio magnitude.

It is clearly of great interest to know how the value of $m_R - m_p$ varies with the type of galaxy, and this is one of the major observational problems. Some measurements of five Sc galaxies suggest that their radio emission, in comparison with their light, is between 1 and 2 magnitudes less than that observed from type Sb, while in the case of the Magellanic Clouds this deficiency has increased to between 3 and 4 magnitudes. At present the value of $m_R - m_p$ for a normal elliptical galaxy is unknown, but the failure to detect any of these galaxies suggests that they radiate weakly at radio wave-lengths in comparison with type Sb.

(b) Peculiar galaxies

In addition to normal galaxies a few radio sources have been identified with peculiar extra-galactic objects. A brief list of these objects is given in Table 2. The first is the well-known source in Cygnus which has been identified by Baade and Minkowski as the collision between two spirals. NGC 4486 is a galaxy in Virgo which has a peculiar jet protruding from the nucleus. NGC 1275 in Perseus has been identified with colliding galaxies; while NGC 5128 is a curious object of which the interpretation is controversial. The table shows that in at least three of these cases the radio emission is greatly enhanced by comparison with a normal Sb galaxy.

The origin of this enhanced radiation is not understood, but it is believed to be due to the collision of gas clouds at high speeds.

Table 2. *Peculiar galaxies*

Object	Radio magnitude m_R (1·9 m)	Photographic magnitude m_{pg}	$m_R - m_{pg}$
Cygnus	+2	+18	−16
NGC 4486	+5	+10	−5
NGC 1275	+7	+13	−6
NGC 5128	+4	+6	−2

5. A COMPARISON OF OUR OWN GALAXY WITH OTHER GALAXIES

Assuming that the distance of NGC 224 (M 31) is $1·5 \times 10^6$ light years, then the value of m_R shown in Table 1 corresponds to an emission of $3·5 \times 10^{20}$ watts ster^{-1} (c/s)$^{-1}$ at 1·9 m. The total radiation from our Galaxy cannot be found precisely since, among other difficulties, it is not known what fraction of the background radiation is of extra-galactic origin. It has been estimated by Hazard, from an integration of the isophotes of the sky at 3 m, that the total radiation is $3·7 \times 10^{20}$ watts ster^{-1} (c/s)$^{-1}$ at 1·9 m, and other estimates range from this value up to about 10^{21} watts ster^{-1} (c/s)$^{-1}$. Since the absolute photographic magnitude of NGC 224 and our own Galaxy are believed to be about equal, it can be tentatively concluded that the ratio of radio to light emission from our own system agrees with that found in the Andromeda nebula; furthermore, within the limits of the present data, this ratio appears to be characteristic of other normal Sb galaxies.

It is also pertinent to inquire whether the distribution of intensity found in our own Galaxy is found in other similar systems. Any attempt to solve this question is beset by the considerable experimental difficulty of obtaining the necessary resolving power. A few attempts have been made to determine the distribution of intensity across NGC 224, but unfortunately the results are not free from difficulties of interpretation. The most significant datum appears to be that obtained by Baldwin with an interferometer using very large aerials. He found that the distribution of intensity at a wave-length of 3·7 m extended over a surprisingly large area, greatly exceeding the region of optical emission; what is even more surprising, he found that the distribution was also extended normal to the plane of the nebula. He interpreted these observations as showing that the nebula contains two separate populations, one distributed in a similar way to the general mass, and one which has approximate spherical symmetry and which reaches to radial distances of the order of 10,000 pc. The latter population contributes about two-thirds of the total radiation.

A comparison of these results with the data presented above suggests that the spherical system in NGC 224 corresponds to the broad distribution observed in our own Galaxy, and that the distribution which is similar to the mass corresponds to the narrow distributions in the Galaxy. Since the distributions across other normal Sb galaxies have not yet been measured we cannot pursue this subject further.

In conclusion, the present data indicate that our own Galaxy, as far as radio emission is concerned, is similar to NGC 224; furthermore, the ratio of radio flux to light appears to be characteristic of normal Sb galaxies.

9. THE RADIO EMISSION FROM THE GALAXY AND THE ANDROMEDA NEBULA

J. E. BALDWIN

Cavendish Laboratory, Cambridge, England

Although the sources responsible for the radio emission from the Galaxy are unknown, it may nevertheless be valuable to make a comparison between the magnitude and distribution of the emission observed in our Galaxy and that of other nebulae. Analyses have already been made to relate the total emitted power from nearby nebulae with that from the Galaxy [1], and by considering the integrated radiation from well-defined clusters attempts were made to extend the comparison to the average emission from fainter nebulae.

The results of this work indicated that there was a reasonably constant relationship between optical and radio magnitudes for nebulae of types Sb and Sc down to magnitude + 10. The extension to fainter magnitudes by considering the integrated radiation of clusters has been shown to be unreliable by the observation of Baldwin and Elsmore of the Perseus cluster [2]; in fact nearly three-quarters of the total emission from this cluster originates in NGC 1275, which is a source of very much greater luminosity than the nearby galaxies and which is now thought to be the result of a collision between a spiral and an elliptical nebula [3]. Owing to the much greater probability of such encounters in clusters, it seems likely that similar difficulties may arise in using the observations of other dense clusters.

Although it has been possible in this way to make certain deductions about the relationship between the optical and radio emission from certain extra-galactic nebulae, a further difficulty arises when a comparison with the Galaxy is attempted. In early investigations, the relative importance of the contributions to the integrated radiation from extra-galactic sources and from sources inside the Galaxy was unknown; it was therefore impossible to predict the intensity which would be expected for other nebulae on the supposition that they were similar to the Galaxy.

The first detailed model of the galactic radio emission was proposed by Westerhout and Oort [4], who showed that the observed contours of brightness of the integrated radiation could be explained on the supposition that the sources responsible for the radio emission were distributed in the same

44

way as the distribution of mass in our Galaxy. They found that the computed values of radio brightness showed good general agreement with the observations except in one important respect: the observed temperatures showed a fairly uniform excess, over those calculated, of about 600° K (at a wave-length of 3·0 m) in all parts of the sky. They suggested that this emission might originate in extra-galactic sources or in some widely extended population of galactic sources.

Recent observations at Cambridge have enabled a new map of the radio isophotes at a wave-length of 3·7 m to be made. Using this new data it has been possible to show conclusively that the greater part of the integrated radio emission is due to a very extended distribution of galactic sources; a similar suggestion has already been made by Shklovsky[5]. The new data not only provide a good estimate of the total power radiated by the Galaxy, but also give a much more reliable model of its spatial distribution.

More detailed measurements have shown that a similar extended distribution of radio emission exists in the Andromeda nebula. Using the new data, a detailed comparison of the radio emission from the Galaxy and the Andromeda nebula has been made.

The new observations of the integrated radiation were made with one section of the Cambridge radio telescope[6]. An examination of the contours of brightness shows that the areas of minimum brightness lie at galactic latitudes of $\pm 45°$ between longitudes 100° and 210°. This result suggests immediately that an appreciable fraction of the radiation at high latitudes may be galactic in origin.

A study was then made of the emission at latitudes greater than 40° where the contribution from the Westerhout and Oort population is less than one-tenth of the total emission. A series of models was constructed each of which was ellipsoidal and of uniform emission per unit volume. An investigation was made of the effects of varying the size and axial ratio of the ellipsoid and also the effect of assuming different values of the extra-galactic radiation. In each case the model was compared with the observations by plotting, as a function of galactic longitude, the intensity at latitudes of +40°, +50°, +60°, and +70°. The figures were also derived for $b = +30°$, but in this case agreement cannot be expected since the Westerhout and Oort distribution here makes a significant contribution to the total radiation. One example of these diagrams is shown in Fig. 1; in this model the extra-galactic radiation was taken to be zero. It can be seen that, apart from the arm of radiation running up to the galactic pole along $l = 0°$ which is a feature seen in many previous surveys, a radius of 16 kpc gives good agreement with the observations.

So far a uniform emission per unit volume has been assumed, but, in the region beyond 10 kpc from the centre of the Galaxy, appreciable variations in the value of the emission per unit volume (σ) would be undetectable. The interpretation of this region is difficult because a spherical shell of sources whose radius is appreciably greater than the distance of the sun from the centre gives a brightness distribution over the sky which is almost isotropic as seen from the earth. It is thus not possible

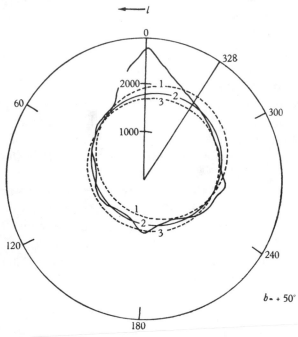

Fig. 1. The variation of brightness temperature with galactic longitude at $b = +50°$. The three circles are the calculated distributions for models having values of R of (1) 12 kpc, (2) 16 kpc, and (3) 20 kpc respectively.

to determine the radial distribution in the outer parts of the sphere; there might, for example, be a significant contribution from extra-galactic sources.

A wide range of satisfactory models may be constructed based on the assumption of different values of the extra-galactic emission; however, it is impossible to account for the observations if this component exceeds 500° K. In every case the best agreement is obtained with a spherical model, but slightly ellipsoidal models would also be permissible. The range of models consistent with the observations is shown in Table 1.

The range of possible models may be narrowed still further if an independent estimate of the extra-galactic emission can be made. It seems probable that this radiation is at least $150°$ K (at a wave-length of $3\cdot7$ m). We may therefore conclude that the greater part of the radio emission from the Galaxy is due to an almost spherical distribution whose radius lies between 11 kpc and 14·5 kpc and that in this region the emission per unit volume is sensibly constant, with a value of $1\cdot8 \times 10^8$ watts ster^{-1} (c/s)$^{-1}$ pc^{-3}.

Table 1

Assumed extra-galactic radiation (° K)	R (kpc)	Emission per unit volume (σ) watts ster^{-1} (c/s)$^{-1}$ pc^{-3}	Axial ratio
0	16	$1\cdot8 \times 10^8$	$> 0\cdot9$
250	13·5	$1\cdot8 \times 10^8$	$> 0\cdot8$
500	11	$1\cdot8 \times 10^8$	$> 0\cdot65$

At galactic latitudes less than $30°$ the contours of brightness are consistent with a distribution of radio emission similar to the distribution of mass in the Galaxy, as suggested by Westerhout and Oort. However, this population contributes only about a fifth of the total radiation from the Galaxy.

These results may now be compared with the observations of the Andromeda nebula. It was shown [7] that in this case also, most of the radiation originates in an approximately spherical region; the observations were not sufficiently accurate to determine precisely the contribution of the Westerhout and Oort population, but it was shown that at least two-thirds of the total emission is due to the spherical population; the proportion could, however, be as great as in the Galaxy. The distribution of brightness across the nebula was shown to be consistent with a spherical distribution having a uniform emission per unit volume and an apparent angular radius of 100′ of arc. Using the most recent determination of the distance of M 31, 610 kpc[8], the actual radius becomes 18 kpc.

In Table 2 a comparison is made between the spherical distribution in the Andromeda nebula and the two limiting models of this component in the Galaxy. The two values of σ quoted for M 31 refer to the cases in which (i) the spherical distribution contributes only two-thirds of the total emission, and (ii) the spherical distribution is responsible for the whole of the radiation.

Table 2

	R (kpc)	Total emission watts ster^{-1} (c/s)$^{-1}$	σ watts ster^{-1} (c/s)$^{-1}$ pc^{-3}
M 31	18	$0\cdot8 \times 10^{21}$	$0\cdot24$–$0\cdot35 \times 10^8$
Galaxy	$\begin{cases} 14\cdot5 \\ 11 \end{cases}$	$2\cdot3 \times 10^{21}$ $1\cdot0 \times 10^{21}$	$1\cdot8 \times 10^8$ $1\cdot8 \times 10^8$

It is interesting to note that, although the limitations of the present analysis lead to an uncertainty of more than a factor of two in the total emission from the Galaxy, the derived emission per unit volume is much more accurately known. The existence of extended distributions of radio emission have been found in both M 31 and the Galaxy; in the case of the Galaxy the radius is appreciably smaller than that in M 31, but the emission per unit volume is about six times greater.

REFERENCES

[1] Hanbury Brown, R. and Hazard, C. *Phil. Mag.* **43**, 137, 1952.
[2] Baldwin, J. E. and Elsmore, B. *Nature,* **173**, 818, 1954.
[3] Baade, W. and Minkowski, R. *Ap. J.* **119**, 215, 1954.
[4] Westerhout, G. and Oort, J. H. *B.A.N.* **11**, 323, 1951.
[5] Shklovsky, I. S. *Ast. Zh.* **29**, 418, 1952.
[6] Ryle, M. and Hewish, A. *Mem. R.A.S.* **67**, 97, 1955.
[7] Baldwin, J. E. *Nature,* **174**, 320, 1954.
[8] Baade, W. Reported in *Sky and Telescope,* **14**, 371, 1955.

Discussion

Schmidt: What is the explanation of the excess radiation at longitude 0°?
Baldwin: We do not know, but it has been observed in several surveys.
Schmidt: Are the minima observed at latitudes $\pm 45°$ clearly represented?
Baldwin: Yes.

10. VARIABLE STARS AND PROBLEMS OF GENERAL STRUCTURE OF GALAXIES

B. V. KUKARKIN

Sternberg Astronomical Institute, Moscow, U.S.S.R.

The general structure of the galaxies is extremely diverse and can be described by Hubble's classification only in the first approximation. In fact, the differences in the distribution of gas and dust in some galaxies which have been established recently, the evident dissimilarity of the intensity of the radio emission in individual galaxies, as well as other facts, suggest definite differences even among galaxies belonging to the same type according to Hubble. Hubble's classification gives a sufficiently exhaustive description of the diversity of the optical forms of galaxies and unites them in a single sequence, which, in the case of spirals, branches into two parallel sequences (ordinary and barred spirals). The interpretation of Hubble's sequence as an evolutionary one became rather popular. In this case usually either the Eo form is considered as the initial stage (Jeans, Hubble) or, conversely, the irregular galaxies and late spirals are considered as the initial stage with the elliptical galaxies being the late stage of evolution (Shapley and many others).

There is no definite evidence in favour of either of these two opposite hypotheses. Moreover, there is no reliable evidence in favour of the statement that Hubble's sequence is an evolutionary one. In fact, every galaxy observed at present, as well as every other cosmic body or system of such bodies, is the result of various initial conditions and of their evolution under the action both of the internal laws and the inevitable influence of the surrounding medium. There are no reasons to suggest (and such reasons cannot exist) that all the galaxies were formed at the same time in some remote epoch and that the great variety in their properties might be explained by only the differences in their initial conditions. At the same time, there are no solid arguments that the Hubble classification of a galaxy is a simple function of its age.

The problem of the origin of the general structure of galaxies could be solved if we had some means of dividing the influence of the initial conditions from that of conditions created during the process of evolution. It is necessary to search for indicators which permit such a division. Apparently, the variable stars which are so typical for the various galaxies and stellar systems may be among such indicators.

The variable stars may be called *marked stars* since nature itself has made them easily distinguishable from the majority of other stars. Simple methods of investigation permit a fine morphological classification of variable stars. Therefore, it is easy to establish the relationship between the general structure of galaxies and the existence of variable stars of different types or of different morphological properties. We are now beginning to discover the first relationships between the general structure of galaxies and the properties of variable stars. The aim of the present short communication is to recall some facts already known and to invite the astronomers of the whole world to carry out detailed studies of the morphological properties of variable stars in different parts of the complex galaxies similar to our Galaxy, in the large Andromeda nebula, and also in other galaxies.

That long-period Cepheids are typical of the flat component of our galaxy and of other spiral galaxies, and the short-period Cepheids, or RR Lyrae-type stars, of the spherical component of our Galaxy, the elliptical galaxies, and the globular clusters, was known long ago. However, a more detailed investigation showed that long-period Cepheids are found even in the spherical component of our Galaxy. At the same time, the morphological properties of these Cepheids were found to be sharply different from those of the Cepheids belonging to the flat component [1,2,3,4,5]. Furthermore, long-period Cepheids in globular clusters are similar to the Cepheids in the spherical component of our Galaxy although they possess some individual properties in every globular cluster. A comparison of the morphological properties of the long-period Cepheids in the flat component of our Galaxy with the properties of the Cepheids in the Magellanic Clouds also shows some fine differences between them in spite of the rough similarity which exists. For example, the change from light curves with a clearly outlined hump on the descending branch to light curves in which a hump on the ascending branch is seen, takes place at periods from 9 to 10 days in long-period Cepheids in the flat component of the Galaxy, whereas in the Magellanic Clouds this change occurs in Cepheids with periods from 11 to 12 days. Such small but definite differences are observed in every isolated stellar system. Perhaps these slight differences depend on small variations in the initial conditions during the process of formation of each stellar system. Besides, in such an extremely complicated system as our Galaxy, the long-period Cepheids of even the flat component are heterogeneous and apparently might be divided into several sequences [6]. The same comments also refer to the short-period Cepheids (RR Lyrae stars).

Of extreme importance in the study of the general structure of galaxies is the question of the zero-point of the period-luminosity relation for Cepheids. Indeed, our ideas about the dimensions, densities, and other important characteristics of different galaxies depend upon the correct answer to this question. Fascinated by the similarity of the period-luminosity relation in the most diverse stellar systems, astronomers sometimes assumed that this relation was the same for the whole universe. However, Baade showed convincingly that the long-period Cepheids in the flat component of our Galaxy are at least one and a half stellar magnitudes brighter than the long-period Cepheids in the spherical component [7]. But it would be premature to support Baade's widely shared opinion that all long-period Cepheids are one and a half or even two magnitudes brighter than the magnitudes obtained by H. Shapley. A detailed study of this question shows that there exists a close relationship between the absolute magnitudes of the Cepheids and their morphological properties. As the data at our disposal testify to an extreme diversity of the morphological properties of variable stars of the same type from one stellar system to another and from one part of a complex system to others, it obviously may be stated with certainty that there are numerous zero-points for the period-luminosity relation for the long-period Cepheids [8].

H. Arp's valuable investigation of the long-period Cepheids in globular clusters [9] speaks most convincingly in favour of the plurality of zero points for the period-luminosity relation for the long-period Cepheids. Further, the study by H. Arp confirms the suggestion of a relationship between the morphological properties of the Cepheids and their absolute magnitudes. The study of these relationships will solve the zero-point problem.

The same may also be said about the RR Lyrae stars. Their median absolute magnitude is constant only for a given stellar system, but is not the same for different systems. As is known, W. Baum's remarkable study of the colour-magnitude diagrams of a number of globular star clusters [10] led him to contradictory results. If the same absolute magnitude is assumed for all the RR Lyrae stars, the absolute magnitudes of the stars in the dwarf sequences in these clusters diverge by more than two magnitudes. On the other hand, if these sequences are superimposed, the absolute magnitudes of the RR Lyrae stars will diverge by more than two stellar magnitudes. Such a contradiction is only apparent. W. Baum solves it with the proposition that we do not find in a globular cluster a main sequence typical of that in the solar neighbourhood, but a sub-dwarf sequence which differs for every cluster. It was definitely suggested in 1949 ([2], p. 181) that in the case of globular clusters we are dealing with sub-dwarf sequences. However, we cannot require tens of thousands of sub-dwarfs in various globular clusters to show systematical differences from one cluster to another while for some reason the properties of the RR Lyrae stars are just the same. But, in reality, the RR Lyrae stars in different globular clusters are different in absolute magnitude. It is interesting to notice that the morphological properties of the RR Lyrae stars change appreciably from one cluster to another. Thus, in the case of the globular clusters, the sub-dwarf sequences should not be superimposed; but, equally, we must not expect that the absolute magnitudes of the RR Lyrae stars should be the same in different clusters. The relationship between the absolute magnitudes of the RR Lyrae stars and their morphological properties should be established. The best method of investigating that question would be a study of the RR Lyrae stars in the globular clusters in the galaxy M 31, when they become accessible.

After a comparison of the existence of variable stars in individual galaxies and other stellar systems, it was stated that there is a definite connexion between the general structure of a given stellar system and the existence in it of certain types of variable stars. Thus, long-period Cepheids are typical of late spirals (the flat component of our Galaxy, M 31, M 33, both Magellanic Clouds); in the spherical component of the spiral systems, the short-period Cepheids are typical, and so forth. Novae are very typical of our Galaxy and M 31, but are much less frequent in the developed spirals like M 33 and in developed barred spirals (Magellanic Clouds). In some galaxies, hot, non-stable (variable) super-giants are typical; in others, we find variable super-giants of type F.

Our Galaxy is extremely rich in variable stars of the Mira Ceti type, while the Magellanic Clouds are poor in them. Mira Ceti type variables in the Clouds are sharply different in their morphological properties from stars of that type in our Galaxy. Mira Ceti variables in the flat component of our Galaxy differ decidedly in their morphological properties from stars of the same type in the spherical component of our Galaxy.

The morphological properties of RR Lyrae type stars in the nucleus of our Galaxy are sharply different from the properties of these stars in the solar surroundings [11].

A great number of such examples might be given. They will considerably increase in the future, as soon as such examples reflect the relationships actually existing in nature. There is every reason to suggest that the existence of variable stars of a given type in a stellar system is connected with the age of that system (see also my communication to the Symposium on non-stable stars [12]).

The promising problem which stands before us is to study in detail the morphological properties of variable stars in different galaxies and to define the relationships outlined between them and the general structure of the galaxies. Undoubtedly, a thorough study of these relationships will be a great help in solving the problems of the evolution of galaxies.

REFERENCES

[1] Kukarkin, B. V. *A.J. U.S.S.R.* **24**, 269, 1947.
[2] Kukarkin, B. V. *The Study of the Structure and Evolution of Stellar Systems.* Moscow, 1949.
[3] Kukarkin, B. V. *Erforschung der Struktur und Entwicklung der Sternsysteme.* Berlin, 1954.
[4] Joy, A. *Ap. J.* **110**, 105, 1949.
[5] Kukarkin, B. V. and Kulikovsky, P. G. *Variable Stars*, **8**, 1, 1951.
[6] Eggen, O. *Ap. J.* **113**, 367, 1951.
[7] Baade, W. *Trans. I.A.U.* **8**, 397, 1954.
[8] Kukarkin, B. V. *A.C. U.S.S.R.* no. 155, 1955.
[9] Arp, H. C. *A.J.* **60**, 1, 1955.
[10] Baum, W. A. *A.J.* **59**, 422, 1954.
[11] Gaposchkin, S. *Variable Stars*, **10**, no. 6, 337, 1956.
[12] Kukarkin, B. V. *Non-Stable Stars* (I.A.U. Symposium no. 3, paper 15, 1955), Cambridge, 1957.

Discussion

Baade: Dr Hubble neither stated nor believed that his classification was an evolutionary scheme. It was simply as a manner of speaking that the E-nebulae were called early. Its interpretation as an evolutionary scheme was a complete misunderstanding.

Secondly, I think the paper by Arp has shown that we do not have a unique period-luminosity relation among the population II Cepheids. Although the data are not sufficient yet to show clearly what this means, there is every evidence that there is really a mixture among the stars we call population II Cepheids. In that mixture we can recognize the RV Tau stars already, but we need more observations. Dr Arp is continuing to get data for more globular clusters, which may clear up this point.

The situation of the zero point of the period-luminosity diagram is not too bad. Both classical and cluster-type variables appear in the Magellanic Clouds at the proper apparent magnitudes. Also, all apparent magnitudes which we can observe in the Andromeda nebula seem to be correct, or at least in agreement with, those in our own system.

Kukarkin: I trust that I have never criticized Hubble for giving an evolutionary meaning to his sequence, but only others who have interpreted his work in that way such as, for instance, Jeans.

At least one point is clearly established: there exist two different zero points for the Cepheids. The one for those which belong to the flat component of our Galaxy is 1·5 magnitudes brighter than that for the spherical component. I agree with Dr Baade that the zero point problems should be stated as problems; they should be investigated; they are not yet solved.

11. THE SPACE DISTRIBUTION OF MIDDLE-AND LATE-TYPE SUPER-GIANTS IN THE REGION OF THE GALACTIC SYSTEM NEAR THE SUN *

WILLIAM P. BIDELMAN

Lick Observatory, University of California, U.S.A.

This paper is concerned with the space distribution in our own Galaxy of the most luminous stars of spectral types A5 and later which have been classified on the basis of low-dispersion slit spectrograms. Seventy stars of luminosity classes Ia and Iab on the Morgan–Keenan system [1] are considered, of which the majority have been previously noted as of high luminosity at the Yerkes, Harvard, Mount Wilson, and Warner and Swasey Observatories. The Cepheid variables have not been considered.

In order to derive the distances of these stars, one needs to know their spectroscopically assigned absolute magnitudes, which are quite uncertain, and also the interstellar absorption suffered by their light. The interstellar absorptions have been estimated in each individual case from the color excesses of neighboring early-type stars of, if possible, similar distances, using the data published recently by W. W. Morgan and his collaborators [2]. These estimated absorptions are of course subject to considerable errors, which may be especially large in the case of the southern stars. The absolute magnitude calibration used is a modification of that given by Keenan and Morgan in the book *Astrophysics* [3]. This modification is suggested by the undoubted presence of a considerable number of these very high luminosity stars in various stellar associations and clusters, especially in those which do not give evidence of being extremely young [4]. As an ex-example of this modification, it may be noted that the most luminous stars of types near G0, such as ρ Cassiopeiae and HR 4337, have been assigned visual absolute magnitude $-8 \cdot 5$. In view of the method of calibration used, the writer's distance scale is tied directly to that used by Morgan, Whitford and Code in their discussion of the space distribution of the blue giants [5].

The positions of the individual middle- and late-type super-giant stars,

* *Contributions from the Lick Observatory*, Ser. II, no. 81.

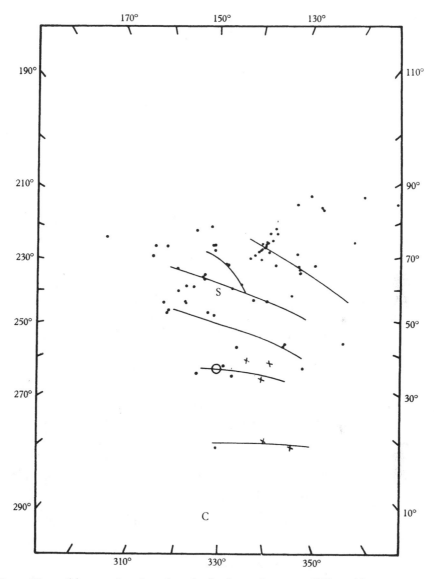

Fig. 1. The positions, projected on the galactic plane, of seventy middle- and late-type stars of high luminosity. Filled circles represent individual super-giant stars, crosses individual early-type stars found by Morgan, Whitford and Code (*Ap. J.* **118**, 318, 1953), and the large open circle an early-type stellar association discovered by Hiltner (*Ap. J.* **120**, 41, 1954). The letter S indicates the position of the sun, C that of the galactic center. The curves nearest the solar position have been drawn in the positions of nearby spiral arms indicated by optical and 21-cm radio data.

projected on the galactic plane, are shown in Fig. 1. The significance of the various symbols used is noted in the figure legend. The curves passing nearest the sun are drawn in the positions of the Perseus arm, the Orion arm, and the first inner arm, which have been previously indicated by optical and 21-cm radio data. The two curves nearer the galactic center have been drawn in to indicate other possible, more distant arms.

In connection with Fig. 1, it must be realized that the data plotted refer to individual stars whose assigned distances probably have uncertainties of the order of 30%. For example, the rather large spread in distance shown by the group of stars near the Perseus arm at galactic longitudes near 105° is undoubtedly largely due to observational error, since it is quite certain that most of these stars are associated with the h and χ Persei clusters, and as such should all actually be plotted at the same distance. Because of the large uncertainties in distance, and also because of the relatively small number of individual stars plotted, it is not to be anticipated that the points in Fig. 1 should actually *define* the complex spiral arms which really exist, even if these high luminosity stars do lie in fact in or near these arms. However, one can legitimately consider the question of whether these middle- and late-type super-giant stars do follow the presently accepted spiral arm pattern which has been *previously* and *independently* established. With due regard to the limitations of the data, it appears to the writer that this question can be answered, in general, in the affirmative. At the same time it should be emphasized that certain of these super-giants, 89 Herculis, for example, are located at rather large distances from the galactic plane, and hence cannot be said to be *closely* associated with any spiral arm of our Galaxy.

Additional radial velocities and three-color observations of these high luminosity stars will undoubtedly permit a considerably more definitive study of the space distribution of these objects.

REFERENCES

[1] Johnson, H. L. and Morgan, W. W. *Ap. J.* **117**, 313, 1953.
[2] Morgan, W. W., Code, A. D. and Whitford, A. E. *Astrophysical Supplements*, **2**, 41, 1955 (no. 14).
[3] Hynek, J. A. (editor). New York (1951).
[4] Bidelman, W. P. *A.J.* **59**, 315, 1954.
[5] Morgan, W. W., Whitford, A. E. and Code, A. D. *Ap. J.* **118**, 318, 1953.

12. THE OBSERVATION OF BLUE GIANT STARS AT GREAT DISTANCES FROM THE SUN

W. W. MORGAN

Yerkes Observatory, Wisconsin, U.S.A.

(Abstract)

A method is described for the discovery and observation of early-type giants which are too faint for ordinary spectroscopic observation in the blue region.

I. INTRODUCTION

There are several limitations to the determination of the space distribution of distant blue giant stars; these are concerned with their discovery and accurate observation for distance. The practical limit for the segregation of the OB Group (Morgan, 1950) [1] with Schmidt cameras of the 24–36-inch type is near 12·5 photographic magnitude. This is also near the practicable limit for the determination of spectroscopic luminosity classes on spectrograms having dispersion of the order of 100 Å/mm.

There are, therefore, two fundamental problems connected with the investigation of fainter and more distant blue giants; there is first the problem of the discovery of a group similar to the OB classification; in the second place, a method must be developed for the determination of individual luminosities and distances for these fainter stars.

The present paper is concerned with the extraction from the general Milky Way population of a residuum similar in characteristics to the OB group. This process is carried out in two approximations. A third approximation for further purification of the residuum is also described.

2. FIRST APPROXIMATION: SEGREGATION FROM UNWIDENED SPECTRA OF VERY LOW DISPERSION

The method developed by Morgan, Meinel and H. M. Johnson (1954) [2] appears to be the most efficient one at the present time for the segregation of a group rich in the percentage of blue giant stars. It depends in principle on the large ultra-violet excess in the radiation of reddened early-type stars as compared to the yellow giants.

The experiments of the above-mentioned investigators were carried out with a small Schmidt camera and an objective prism giving a mean dispersion of around 30,000 Å/mm. The principle, however, can be generalized for use with large Schmidt cameras, combined with objective prisms having refractive angles of around $\frac{1}{3}°$. The important consideration is that the length of the spectra from the red to the ultra-violet shall be of the order of 10 or 15 times the diameter of faint focal images.

Experiments with a small 6-inch prism of this type have been carried out by D. Schulte with the Yerkes 24-inch reflector. Although the angular field in this case is small, he has shown that a residuum very rich in early-type stars can be separated from the general Milky Way population to around the 14th photographic magnitude. A similar prism, when combined with one of the 24–36-inch Schmidt cameras should make possible a most efficient first approximation for the extraction of early-type stars.

The residuum obtained in this manner consists of 30–80 % of blue giants; the remainder are principally G 5–G 8 giants and main-sequence F–G stars.

3. SECOND APPROXIMATION: PURIFICATION OF RESIDUUM BY THREE-COLOR PHOTOELECTRIC PHOTOMETRY

Fig. 1 illustrates the location of reddened stars of early type on the (U, B, V) color system of Johnson and Morgan. The observations are by H. L. Johnson and D. L. Harris III; most of them have been published already; I am indebted to Dr Johnson for the use of some unpublished observations. The position of the normal main sequence is illustrated; also, the location of unreddened stars of early O-type and of class B 1 V, respectively, are shown by the two large circles. A 'reddening-line' having a slope of 0·70 has been drawn through the former; the reddening-line drawn through the latter has a slope of 0·80. The observed variation in the slope of the reddening-line probably lies between these extremes [3].

The band between the two reddening-lines therefore includes the reddening-paths for stars of spectral types O 5–B 1; in addition, B 2Ia, B 2Ib and B 3Ia stars are also included.

Fig. 1 also shows the position of the G 5–G 8 giants and the F 8–G 5 main sequence. The former are located in the neighborhood of (B–V = +0·9; U–B = +0·7); the G dwarfs are in the neighborhood of (B–V = +0·6; U–B = +0·05). It can be seen, therefore, that these two groups can be distinguished from the blue giants with which they are mixed in the first approximation by comparison of the (U–B) index. In addition, blue stars of spectral types B 2 and later can be distinguished, since they lie

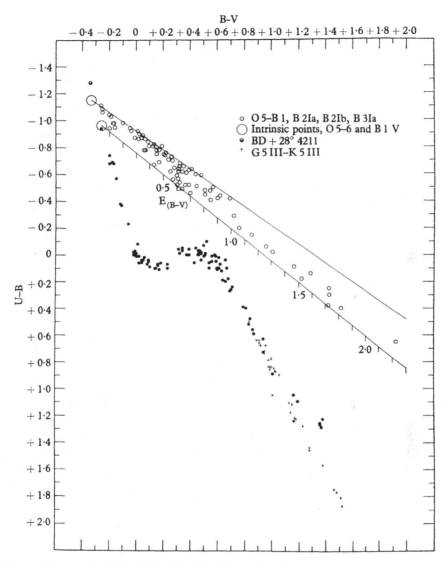

Fig. 1. Diagram showing position of reddened stars of early type. Solid dots give position of unreddened main sequence from B 1–K 7. The slope of the reddening-line through the O 5–6 intrinsic point is 0·70; that through the B 1 V intrinsic point is 0·80. These two values lie near the extremes found for the regional variation of the reddening law as found by Johnson and Morgan.

below the reddening band illustrated; the stars of spectral types B 2 and B 3 of highest luminosity are included in the band.

The second approximation, therefore, allows the yellow giants and main-sequence stars, and the blue stars of later B subdivisions, to be separated from the first residuum and discarded.

We have, at this stage, a group which approximates closely the OB group as defined by Morgan from slit spectrograms. At the high-temperature end, the two groups are almost identical; the principal difference is in the case of the B 5–A 5 class I stars, which are situated below the band illustrated in Fig. 2. Some main-sequence Be stars which have unusually strong ultra-violet radiation may also be included.

There is an additional characteristic of stars within the reddening band which is of considerable importance for investigations of galactic structure: the range in intrinsic color on the (B–V) system is small. The intrinsic color of the O 5–6 stars is about -0.33 magnitude; the intrinsic color at B 1 V is about -0.26 magnitude. The intrinsic colors of B 2Ia–B 3Ia stars are probably somewhat less blue; however, even with this consideration, color excesses of good accuracy can be derived from a mean intrinsic color for the group; for the present, a round value for intrinsic (B–V) of -0.30 magnitude has been adopted; the definitive value will probably lie close to $(B–V) = -0.28$ magnitude.

With a factor of around 3 to transform color excesses to total visual absorptions, the latter should have probable errors of the order of ± 0.1 magnitude. Thus, without any knowledge of spectral types or luminosity classes, total absorptions in the line of sight of fairly high accuracy can be derived.

The circumstances are not so favorable in the case of the absolute magnitudes of the stars included in the reddening band; the range in visual luminosity is from around $-2^{m}.5$ to around $-7^{m}.5$ on the V-system. The situation is, therefore, that while accurate interstellar absorptions can be derived for individual stars located in the reddening band illustrated in Fig. 1, individual luminosities cannot be safely derived.

4. THIRD APPROXIMATION: PURIFICATION OF SECOND RESIDUUM FROM LOW-DISPERSION SPECTROGRAMS IN THE RED REGION

The make-up of the second residuum is as follows: (1) O–B 1 stars, together with B 2–3 super-giants; (2) under-luminous blue stars similar to BD $+28°$ 4211; (3) white dwarfs; and (4) the possibility of the presence of T Tauri stars of type K. Of the various groups, all but group (1) will be relatively rare—as long as observations are confined to the region of the galactic equator. Since the method is intended for application to stars of the 12th magnitude and fainter, the white dwarfs can be immediately recognized from their low color excesses, in comparison with stars of group (1) in the vicinity; it is probable that the sub-luminous blue stars in group (2) can also be recognized in the same manner; however, in this case and also in the case of group (4), a further purification can be effected by the obtaining of low-dispersion slit spectra in the red region. For reddened early-type stars located at distances greater than 2 kpc, the interstellar absorption at λ 6284 is easily observable on slit spectra with a dispersion of 350 Å/mm. Such spectrograms are obtainable with ease with reflectors of the second category of size (the McDonald 82-inch reflector).

In the case of the under-luminous blue stars and the T Tauri dwarfs, the strong interstellar absorption would be weak or absent; these objects could then be separated from the residuum and discarded.

After the third approximation, the residuum is a well-defined group of O–B 1 stars, B 2–B 3 super-giants, and a certain number of B 2e stars having abnormally bright ultra-violet radiation. For these stars we will then have an accurate determination of their color excesses, and apparent magnitudes corrected for interstellar absorption.

The problem of the derivation of distance can be solved fairly simply in the case where a rich segment of a distant spiral arm appears to have been observed. The only additional information necessary is the calibration of the bright end of the luminosity function for the residuum after the third approximation.

REFERENCES

[1] Morgan, W. W. *Publ. Obs. Univ. Michigan*, x, 33, 1950.
[2] Morgan, W. W., Meinel, A. B. and Johnson, H. M. *Ap. J.* **120**, 506, 1954.
[3] Johnson, H. L. and Morgan, W. W. *Ap. J.* **122**, 142, 1955.

Discussion

Oort: I was intrigued by Dr Morgan's remark about the difference in slope of the U–B vs B–V color excess lines in different regions. Can Dr Morgan give some more information on this point?

Morgan: In an investigation of the reddening law in the direction of the great rift in Cygnus, based on photo-electric colors by H. L. Johnson, a comparison of an ultra-violet-blue plot, such as I have shown here, for moderately and extremely reddened O-stars in the rift with one for those in galactic longitudes between 90° and 180° gives quite strong evidence that there is a small but definite difference in slope. The difference in slope is about 0·1 (from 0·72 to

0·8) in the direction that the ultra-violet is more strongly absorbed in Cygnus. A tentative suggestion is that in Cygnus the O-stars are quite far behind the rift so that one is observing a sample of interstellar material which is at a great distance from early-type stars while, within 2000 pc from the sun in longitudes 90° to 180°, the major part of the absorbing material seems in general to be in the neighborhood of the O-stars. It seems to be a general rule that in this direction the space between the O-stars and the sun is probably fairly transparent except for the local region of high density.

Oort: There is a contradiction between these results and those of Mlle Divan.

Lindblad: Dr Schalén has found good evidence of a variation of space absorption in different regions.

13. THE SPATIAL DISTRIBUTION OF NOVAE AND SUPER-NOVAE

I. M. KOPYLOV

Crimean Astrophysical Observatory, Simeis, Crimea, U.S.S.R.

(Abstract)

In 1945 McLaughlin found that novae form a comparatively flat sub-system in the Galaxy with an appreciable concentration towards the centre. B. V. Kukarkin came to the same conclusion in 1949. However, it was found later that this result, which contradicts our ideas of galactic structure, was a result of poor absolute magnitudes and the consequent underestimate of the distances of the novae.

Therefore, we decided to redetermine the characteristics of the novae sub-system. Including only novae within $1 \cdot 5$ kpc of the sun, allowing for interstellar absorption, and assuming that the absolute magnitude of the nova at maximum is given by: $M = -13 \cdot 7 + 3^{\mathrm{m}} \cdot 5 \log t_3$, where t_3 is the duration of the first three-magnitude decline, we find

$$\frac{\partial \log D}{\partial R} = -0 \cdot 22 \pm 0 \cdot 01,$$

$$\frac{\partial \log D}{\partial z} = -2 \cdot 39 \pm 0 \cdot 11.$$

The degree of galactic concentration is $\beta = 182 \pm 10$ pc. Thus, the novae form a typical, intermediate sub-system in the Galaxy.

The above results refer only to the region around the sun. To extend our results to other parts of the Galaxy, we can compare our Galaxy with the data obtained by Hubble and Mayall in M 31. The distance modulus of M 31, corrected for absorption, is $m_0 - M = 22^{\mathrm{m}} \cdot 9$. For $R < 2 \cdot 5$ kpc, the surface density of the novae increases more rapidly than for $R > 3$ kpc, outlining a nucleus of a considerably increased density. The radius $R = 1 \cdot 6$ kpc can be considered as the boundary of the central nucleus of the novae sub-system at which $\dfrac{\partial \log D}{\partial R}$ reaches its maximum value of $-1 \cdot 12$.

The considerable apparent concentration of novae toward the centre in M 31 may be explained by a nucleus of increased density; at $R \sim 7$–8 kpc, the parameters of the novae sub-system are also typical for an intermediate sub-system.

Additional evidence is given by the fact that both elliptical galaxies and Sc galaxies, typical of the spherical and flat sub-systems, respectively, have very few novae.

The apparent distribution of super-novae in other galaxies suggests that there are two types of these objects. About 20 % are found in the nucleus; the remainder are found far from the nucleus or in spiral arms. Moreover, the super-novae in the nuclei are about 4 magnitudes brighter and have different light curves from those found at greater distances from the centre.

By the identification of some discrete radio-emission sources in the Galaxy with remnants of super-novae, I. S. Shklovsky has found that there have been at least 9 super-novae outbursts in the vicinity of the sun during the last 2000 years. The apparent distribution of super-novae in other galaxies and the low galactic latitude of the super-novae in our Galaxy testify that super-novae belong to comparatively flat sub-systems. Near the sun we estimate:

$$\frac{\partial \log D}{\partial z} = -4 \cdot 9 \quad \frac{\partial \log D}{\partial R} = -0 \cdot 16.$$

14. SPIRAL STRUCTURE AND THE
ROTATION OF GALAXIES

B. A. VORONTSOV-VELYAMINOV

Sternberg Astronomical Institute, Moscow, U.S.S.R.

(Abstract)

There is no general agreement about either the location or the direction of winding of the spiral arms in our Galaxy. Since the measurement of the distances of gaseous nebulae is based on their association with hot stars, the locations of arms based on these objects are not independent.

We have found [1] that the location of most OB-stars and Cepheids does not agree with the concept of trailing arms. Morgan, Whitford, and Code [2] have found the expected trailing of the arms from their twenty-eight aggregates, but they have no data for the southern hemisphere, while, in addition, but a small percentage of the known OB-stars are represented in these 28 aggregates. The 21-cm radio observations have made the situation more confused. Moreover, these observations contradict a number of observational data from other spiral galaxies.

The arms in all galaxies are well represented by a logarithmic spiral $R = a.e^{b\phi}$ with a characteristic angle, $\mu = \text{arc ctg } b$, which ranges from $54°$ to $86°$. At best, the radio observations could refer to a spiral with $\mu = 86°$, but this would make our Galaxy a rare exception to the general run of galaxies. Only 2 or 3 inner arms have been observed. Yet if $\mu = 86°$, at least 8–10 twines must exist.

None of the rotation laws so far suggested maintains logarithmic spiral arms, yet various arguments point to spiral arms as rather stable structures. We conclude that solid rotation is the only possible rotation law for spirals. Since curved bars are unknown, barred spirals must also rotate as solid bodies; Mayall [3] states that observations do not rule out solid-body rotation for the main body of M 31. A sharp change of angular velocity is seen just at the place where the spiral structure in M 31 and M 33 is vanishing. This confirms our conclusion that differential angular rotation and spiral structure are incompatible. Since the arms are not formed by rotation, they may either lead or trail. Conversely, even if the spiral arms are found to wind the same way in all galaxies, this cannot be due to rotational forces.

From optical observations in a sphere of 1·5 kpc around the sun, it has been established only that $\omega(R)$, the angular velocity, decreases from $R \sim 7$ kpc outward and for $R < 7$ kpc, $\Theta(R)$, the linear rotational velocity, begins to decrease inward. Hence, the evident spiral structure must vanish within $R \sim 7$ kpc and by $R > 8$ kpc, even its traces must disappear. Several observations confirm this. The 21-cm data have been misinterpreted; the velocities cannot be converted to distances so simply.

The dynamical models of the spiral galaxies must be revised. We must, possibly, remember the viscosity of the flat galactic layer; it contains much gas in which the free path of atoms is much shorter than the diameter of the Galaxy; it is not a vacuum.

REFERENCES

[1] Vorontsov-Velyaminov, B. A. *A.J. U.S.S.R.*, **30**, 37, 1953.
[2] Morgan, W. W., Whitford, A. E. and Code, A. D. *Ap. J.* **118**, 318, 1953.
[3] Mayall, N. U. *Publ. Obs. Univ. Michigan*, **10**, 19, 1951.

15. ON THE STRUCTURE OF GALAXIES ACCORDING TO PHOTOMETRIC STUDIES

D. E. SHCHEGOLEV

Pulkovo Observatory, U.S.S.R.

(Abstract)

The colours and brightnesses of various details about 12″ or larger in diameter were measured on photographs taken in the two wave-lengths 4360 Å and 6100 Å, of the spiral galaxies: NGC 224, 598, 3031, 3623, 3627, 3628, 4192, 4217, 4254, 4258, 4321, 4565, 4631, 5194, and 5457. The plates were taken with the 500 mm meniscus astrograph at the Kazakhstan High Altitude Observatory near Alma-Ata. Extra-focal stellar images taken with a 'two storied plate holder' were used for standardizing.

These studies show that all nebulae have a bright continuous background which decreases in intensity with the increase of the distance from the nucleus. In Sc galaxies in which the arms are well developed, the arm itself always contributes less than half of the yellow light. In blue light, the arm emission can equal the background, but it exceeds it only in the outermost regions of the nebula. The same is true to an even greater degree in the Sb galaxies.

The colour of the background, which is similar in all regions and nearly that of the nucleus, coincides with the predicted colour for stars in the spherical sub-system (type II). The spiral arms, if the background is excluded, are much bluer (up to $-0^{m}\cdot 1$), corresponding to the colour of stars belonging to flat sub-systems (type I). This colour is caused, to a considerable extent, by the numerous blue condensations located in these arms. The brightest of them are always the bluest ones. Many are similar to O-associations in brightness (up to 20 m/sq sec).

Thus photometric measures indicate some structural features common to all spiral galaxies which have already been found for our Galaxy by other methods:

(1) The largest part of the light and mass is contributed by the yellow stars of the spherical and intermediate sub-systems.

(2) Spiral arms are formed by blue stars which belong to the flat sub-system.

(3) The concentration of the absorbing matter toward the galactic

plane is smaller than for type I stars but larger than for the background stars.

It is usually assumed that the darker, redder side of a galaxy is the more distant one. However, if the largest part of the light in the spiral galaxies is contributed by stars belonging to the spherical and intermediate sub-systems located outside the layer of the absorbing matter, the more distant part of the Galaxy will appear brighter, while the darker side, with a well-outlined spiral structure and which absorbs the light of the majority of stars to a considerable extent will be the nearer side. In this case, the spiral arms will be 'trailing'.

16. COMPARISON OF THE GALACTIC SYSTEM WITH OTHER STELLAR SYSTEMS

J. H. OORT

University Observatory, Leiden, Netherlands

A comparison of the Galactic System with other stellar systems might be based on:

(*a*) the regular aspect (the population II);

(*b*) what we might call the semi-chaotic aspect, presented by the gas and the other population I objects;

(*c*) the spherical corona of continuous radio-emission.

As far as (*a*) is concerned we are not yet in a position to say very much. The only type of objects of extreme population II that we have recognized with certainty in the Galactic System as well as in other comparable systems are the globular clusters. They are very incompletely known in the Galaxy, perhaps somewhat more completely in the Andromeda nebula, but few systematic studies have been made in other systems. So far as we can tell, the Andromeda nebula and the Galactic System are comparable in the number and general distribution of globular clusters.

The bright yellow giants that show such a regular distribution in the disk of the Andromeda nebula have still hardly been recognized in our own Galaxy.

There is, however, one other type of information concerning the general distribution of population II in the Galactic System, namely through measures of rotation. These yield information on the mass distribution. It appears likely that most of the mass consists of population II objects. Concerning the mass distribution, however, the information on *other* stellar systems is still fragmentary. The most we can say at present is that, up to about 7 kpc from the centre, there seems to be reasonable agreement between the rotation curve of the Andromeda nebula and that of the Galactic System (see Fig. 1). The absolute velocities may be a little higher in the Andromeda nebula.* The rotation curve observed in M 33 appears to be considerably different.

* *Note added November 1956.* Recent measures of the 21-cm radiation from the Andromeda nebula have shown that there is no observable difference between the rotational velocities of the south-preceding and the north-following half of the nebula, and that the velocity remains practically constant from 8 to 22 kpc distance from the centre. In the Galactic System, on the other hand, there appears to be a considerable drop in rotational velocity beyond 7 kpc.

The evidence available points strongly to an Sb type for the Galactic System. If it were of type Sa, the strong central bulge would certainly be expected to show up very clearly in the general region surrounding the centre, which it does not.

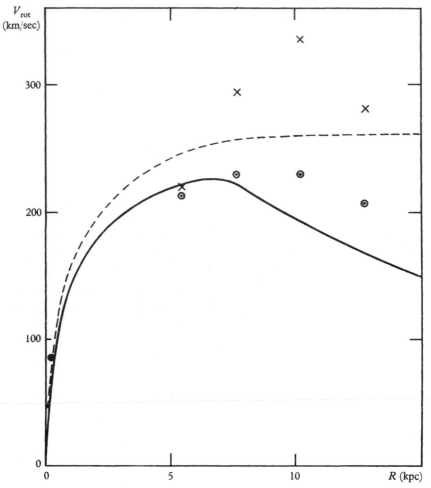

Fig. 1. Comparison of rotation of Andromeda nebula (dashed curve) with that of the Galactic System (full curve). Circles refer to the south-preceding, crosses to north-following side. They were read from smooth curves based on Mayall's and Humason's measures of emission patches. The filled circle at $R=250$ pc refers to Babcock's observations of the inclination of absorption lines in the nuclear part.

As regards (c), observations are available for the Galactic System and the Andromeda nebula only. They indicate a similarity of the two systems in this most interesting and still barely understood feature. But observations are still uncertain in both cases.

70

By far the best material for a comparison of the Galactic System with other systems is furnished by (*b*), i.e. by the distribution of gas and super-giant stars. From the data on the distribution of gas over a large part of the system, as has been described by van de Hulst, we can infer directly that the Milky Way System must be an intermediate-type spiral. Outside the part within 3 kpc from the centre, which has not yet been studied satisfactorily, we find that a radius vector cuts four major arms plus a thinner outer arm. The major arms are spaced at distances of roughly 2·0 kpc and are thus wound moderately closely, such as we would find in an Sb spiral, possibly a somewhat early Sb. The outermost arm seems to deviate more strongly from a circle (cf. Fig. 1, p. 18, in the article by van de Hulst).

This picture shows much resemblance to that which has been described by Baade for the Andromeda nebula. According to his description there are beyond 3 kpc six spiral arms, of which the outermost may be no more than a few dispersed groupings of early-type stars with little interstellar material. On the basis of his new distance determination he estimates the average separation between the arms as about 3 kpc, against 2 kpc in the Galactic System.

In the region closer to the nucleus, however, we note several differences. At first sight it even looks as if the two pictures are to some extent comple-mentary. In the Andromeda nebula the arms appear to grow heavier when we proceed inwards, and they extend down to the very nucleus. In the Galactic System, on the other hand, the density of hydrogen atoms is smaller in the central parts than in our neighbourhood, and we observe only little sign of spiral structure within 3 kpc from the centre. The hydrogen clouds observed in this part of the system appear to move with quite high random motions.

Does this mean that the two systems are radically different from each other? A possible alternative is that the hydrogen in the inner parts of the systems is largely molecular, so that it cannot be directly observed. The high-velocity hydrogen atoms that we do observe would then be only a small admixture.

It remains an intriguing problem to find out what causes these high velocities, and also why it is that in the very dense inner arms of the Andromeda nebula, that are full of heavy dust clouds, no high-luminosity early-type stars are visible.

An interesting set of data which is obtained by the study of the hydrogen distribution is the measure in which the hydrogen layer deviates from a plane. The deviations of the centre of the main layer from the average galactic plane as defined by the 21-cm measures are shown in Fig. 2. It will be seen that in the part within $R = 8$ kpc the deviations are everywhere surprisingly small. It is only in the outer parts that sensible deviations occur. These are still small compared to the distances from the centre. They have, however, an outspoken systematic character.

While the study of the Andromeda nebula and other spirals will help us to obtain a better picture of the general structure of the Galactic System, the study of spiral structure in the Galactic System is bound to become of considerable importance for the understanding of the mechanism that causes the spiral arms.

It has become increasingly probable in recent years that spiral structure is closely related to the gaseous constituent of a system: only systems containing a fair proportion of interstellar gas seem to be capable of developing spiral arms. It is indeed not improbable that the spiral arms *consist* largely of gas.

It is a remarkable fact that practically *all* systems that do contain gas, and that have a sufficiently short time of revolution to have made possible some measure of systematic arrangement of matter, show spiral structure.

So far, however, we have only the vaguest ideas, if any at all, of how a large-scale spiral structure, extending over an entire stellar system, can have originated, and it is even far from clear how, if once originated, it can subsist during several revolutions.

The study of spiral arms in the Galactic System will certainly help to bring us to a better formulation of these problems. The observations in our own Galaxy reveal several features that cannot be observed with comparable precision in other systems, such as the thickness of the layer to which the spiral pattern is confined, and systematic deviations from the average plane of the spiral. Further, we have in the Galactic System a far better knowledge of the gravitational field in which the arms move, while we can also hope, in this case, to obtain an approximate picture of the *stellar* population of a spiral arm. Finally, and perhaps most important, we can study in fair detail the systematic and random motions of the gas clouds that form the basic pattern of the arms, and we may thus be able in a near future to obtain a more complete picture of how the spiral structure might be kept in existence during 10 or 20 revolutions of the system.

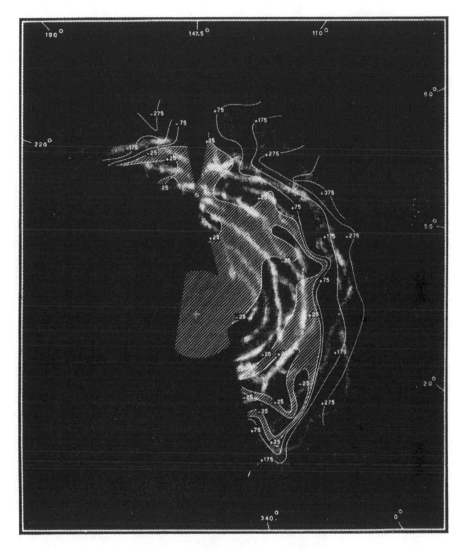

Fig. 2. Deviations of the hydrogen layer from the average galactic plane. This average plane is inclined 1°·5 to the standard galactic plane. In the hatched part the deviations of the centre of the hydrogen layer from the average plane are less than 25 pc. The curves are contours of equal distance from the average plane; the numbers indicate the distances in parsecs.

Facing p. 72

Printed in the United States
By Bookmasters